Origins
The Study Guide

The Origin of Matter, Space, Time, and Life.

Section 1 of 3

by Dr. Troy E. Lawrence
email address: Lawrence@creationministry.org
Send donations to: www.creationministry.org

Edited by George Macias

Published by Troy Lawrence Publishing
ISBN: 978-1943185016

Printed in the United States of America.

All rights reserved. No part of this publication may be reproduced, stored in a retrieval system, or transmitted in any form or by any means—for example, electronic, photocopy, recording—without the prior written permission of the publisher. The only exception is brief quotations in printed reviews. Copyrighted ©.

All Scripture quotations, unless otherwise indicated, are taken from the Holy Bible, New American Standard Bible®, copyright © 1960, 1962, 1963, 1968, 1971–75, 1977, 1995 by The Lockman Foundation. Used by permission.

Table of Contents

Preface 4
Introduction 4

Chapter   **Page**

Section I
How Young is the Earth?

Ch	Title	Page
1	Gravity	7
2	The Effects of Weaker Gravity on Life	26
3	The Canopy of Salt Water	35
4	Climate	48
5	Oxygen Concentration	55
6	Land Was More Plentiful in the Past	61
7	Meteors, Asteroids, and Comets	64
8	Earth's Spin at Origins	70
9	The Flood	71
10	No Deserts Before the Flood	78
11	When and What Caused the Polar Ice Caps and the Ice Age?	86

Preface

This study guide will help explain the origins of dinosaurs, humans, the earth, the sun, the moon, stars, galaxies, ice ages, polar ice caps, deserts, layers of sedimentation, petrification, fossils, vast oil reserves, and the oceans. In addition, this study guide explores the difference between evolution (macroevolution) and adaptation (microevolution) and the difference between evolution and creation. You will learn how dinosaurs became so large and why they are not visible today and what caused the dinosaurs to be extinct. We will cover such topics as whether it is probable for humans to live 900+ years of age and whether humans evolved from primates (monkeys). Did dinosaurs and humans live together on the earth? How old is the earth? What is the origin of all living creatures? These questions and many more will be answered in this study guide.

The more knowledge and truth we possess, the less amount of faith is required to accept the testimony within the Bible. Accepting the records in the Genesis creation account actually becomes easier as knowledge is increased. This is in accord with Romans 10:17, "faith comes by hearing, and hearing by the Word of God." Thus, as you increase in your knowledge of the Scriptures, then your faith grows. The results from believers reading *Origins* is that their knowledge increases, which increases their faith, which leads to them to reading the Bible more frequently and witnessing more. The result from nonbelievers reading *Origins* is that they see that there is a superior interpretation of the observable evidence that is in harmony with science. After reading Origins, some nonbelievers have reportedly rejected a major tenant of evolution, acknowledged that mutations enhance DNA to eventually form new functions and new kinds of creatures, and conceded that there must be some Intelligent Designer out there that they need to search out.

This study guide should not be read by itself; it should be accompanied by the book *Origins* because it contains more detail, footnotes, and additional evidence to solidify each point.

Introduction:
The Premise of *Origins*

One of the major components of this study guide is an explanation of how changes to Earth's gravity, oxygen concentration, rotational velocity, temperature, and the loss of a canopy of salt water that surrounded our atmosphere have adversely affected all living creatures on the earth—resulting in a severe reduction in the length of life, the size of life forms, and in the earth having four seasons, polar ice caps, deserts, and more.

The Motivation for *Origins*

There were several motivating factors that caused me to write *Origins*. The primary factor is that the reigning accepted hypothesis, evolution, is at odds with the Bible. Both cannot be correct since they make polar opposite statements on the beginning of life. Secondarily, there are too many people that have no idea what to think or believe regarding the origins of life and all things.

Origins is such an important topic because how people view the origins of life shapes their entire outlook on all faiths, hope, security, the source of truth, and the purpose for life. There are basically two views—and it's not science versus God, it is evolutionary scientists versus science and God—regarding the origins of life:

(a) The Big Bang initiated all processes and laws, and all life forms evolved from a single-celled organism that spontaneously began from a primordial complex chemical mixture into the complexities of life we see today by explainable natural processes in nature.

(b) God created everything by natural and supernatural processes in six rotations of the earth.

All of public academia, from elementary school age to graduate level education, teaches evolution. And almost all media teaches evolution. This teaching is so ubiquitous that even 90% of seminary schools and most churches teach an old Earth time line to coincide with the evolutionary model. Some churches say that evolution is truth, but God is behind it all. The overwhelming majority of people on earth say that the Bible requires too big of a leap of faith to believe that it's 100% truth; it is just a good moral book. Even the Pope has accepted the old Earth hypothesis and evolution. When broad is the way and many follow, that should give us pause (Matt. 7:13–14).

The serious implications of the origins of life are striking. Start with the wrong premise, and conclusions built on that premise will be in error. If evolution is correct, then the Bible is not from God, but from men and filled with errors. If the Bible's account of creation is correct, then evolution is error. There is no middle ground, since the two views are so opposite. There is no room for compromise. If one verse in the Bible is wrong, then it's not a book from an all-powerful God; it's a book from fallible humans. Frankly, it would need to be burned. But if the Genesis creation account is correct, then evolutionary teaching and doctrine need to be exposed and stopped.

Why does it have to be so cut and dry? Why can't there be tolerance of both views? Because the Bible takes a very hard line and proclaims that there is no middle ground. Jesus said, "He who is not with Me is against Me; and he who does not gather with Me scatters" (Matthew 12:30). Therefore, there is a lot at stake at the very beginning of the origins of life because the Bible declares God cannot lie (Titus 1:2), so if the Bible is in error, then God is a liar and the Bible is made up by man.

An honest scientist would say that evolution and the Big Bang are accepted hypotheses, but invariably when professors lecture, they say evolution is fact. The prevailing, leading, and accepted view is that evolution and the Big Bang are not just possible hypotheses that are further being explored as good theories: they are facts. The implication is that if the Bible's creation account has one error, it is terminal. But on an individual level, people are less likely to read, study, or memorize a book they believe contains errors. Thus, they are less knowledgeable on how to be obedient to God, and by default, more prone to erroneous activity, thought, and doctrine. And the subsequent result is that they are less likely to share truth with others because no one wants to approach an atheist and say, "Believe in my all powerful God, though He wrote errors in Genesis creation."

Are there holes in the evolutionary hypothesis? Is evolution fact? If there are holes in the evolutionary model, there is too much ignorance amongst Christians to know them and too much ignorance about what the Bible teaches for them to defend the Bible. Most people don't know what to think or believe regarding the origins of life. They merely accept whatever they are told. This bothers me for two reasons.

Reason No. 1: The Bible declares that God wrote the Bible through mankind (II Timothy *3:16)*. Therefore, if there is one error in the original manuscripts, for example, in such things as the Genesis account of creation, then the whole Bible can't be infallible and be written by God. It should be viewed simply as a good book along with any other good book. But if the Genesis account of creation is the truth, then indeed the Bible is the Word of God, and evolution and the singularity of the Big Bang may represent the fulfillment of I Timothy 4:1 "But the Spirit explicitly says that in later times some will fall away from the faith, paying attention to deceitful spirits and doctrines of demons."

Reason No. 2: II Timothy 4:1 says,

> I solemnly charge you in the presence of God and of Christ Jesus, who is to judge the living and the dead, and by His appearing and His kingdom: <u>preach the word; be ready in season and out of season</u>; reprove, rebuke, exhort, with great patience and instruction.

Ignorance amongst Christians on the subject of origins is disobedience to the Word of God and is sinful. Christians are at war spiritually for the truth, and the Word of God is the sword (Ephesians 6:12–

17). To be ignorant of what the Bible says about the beginning is an act of willful negligence. To claim to be a Christian and accept evolution as truth and to believe the Bible to be in error is an act of spiritual treason. There is no neutral ground; Christ said you are either for me or against me.

Of those that believe in God, there are too many that just don't have a clue about what happened in the beginning. And this certainly renders them catatonic when an opportunity comes for them to discuss the difference from the Genesis account of creation and evolution. Imagine how ineffective a Christian becomes at witnessing when they tell others to believe in Jesus and His Word, yet they themselves don't even believe Jesus' Word as truth. Why would anyone believe in a God of the Bible that doesn't speak truth regarding the origins of life? They wouldn't.

The battle for the beginning is extremely critical. And for this reason, Satan attempts to separate humans from the Word of God before tempting. Consider Adam and Eve, Satan said, "Indeed has God said," and then he tempted them (Genesis 3). And after Jesus was baptized, anointed by the Holy Spirit, and heralded by the Father with, "This is My beloved Son," then Satan attacked the Word of God by saying, "If You are the Son of God" and then tempted Him (Matt 3:17– 4:9). Therefore, we may conclude that Satan understands the best way to tempt someone to sin is to first attack the Word and separate people from the Word, and then they are easy prey to lead into sin. And there is no better way to do this than to start with *In the beginning*. We have seen the world reject God's creation account and accept man's hypothesis of the beginning and the exponential rise of sin as a result.

Let's journey together through the sciences, logic, and the Bible to determine what is fact or fiction, what is truth or error, and what is from God or from man.

Group Discussion:

1. Have you heard evolutionists proclaim their view as fact? How did that impact you?

2. As a result of the dominant view of evolution, has this limited the amount of time you have read the Bible or shared the gospel?

3. Have you ever felt defenseless to defend creation? How did that affect you?

4. If this book proves that the Genesis creation account is true and that the earth and universe are young, how will that effect your reading and sharing of the Bible?

Origins

Section I
How Young is the Earth?
Chapter 1
Gravity

Was gravity on the earth the same 6 to 10 millennia ago as it is today? No.
Did dinosaurs live in the same gravitation force that we do today? No.
Did the first man and woman live in the same gravitational force that we do today? No.
A foundational premise of *Origins* is that the gravity of the earth is stronger today than in prehistoric times. This stronger gravity has altered life by reducing the length of life and the size of life on earth. And this book will set out to prove it.

Gravity is defined as the gravitational attraction of the mass of an object for bodies at or near its surface. Gravity is based on the mass of the planet. The greater the mass, the greater the gravitational force affects the space and time around an object that then alters the natural path of the object.

There are several methods by which the earth's gravitational force increases over time. One is by direct accumulation of mass. This is by space dust and asteroid impacts. Some scientists estimate that 40,000 tons is added to the earth each year. This seems like a lot of weight, but compared to the large mass of the earth, this is negligible.

When space had many more asteroids floating around and the earth was inundated by asteroids with much greater size and frequency in the past, then that would have increased the rate of accumulation, adding to the earth's mass. Hydrological cycles, tectonic plates, forest growth, water, and erosion have covered up most historic impact craters on earth. But NASA and other scientists have discovered that the earth was impacted by asteroids with more frequency and with larger intensity in distant past millennia. Given the moon's vast amount of crater history and scientific findings about hidden alleged "dinosaur killer" impact craters on the earth, it's conceivable that in past millennia, asteroids and comets hitting the earth were numerous enough to effectively increase earth's gravity by two means:

1. Direct accumulation, adding mass to the earth, which increases earth's gravity.
2. Indirectly by asteroid impacts slowing earth's rotational velocity, which decreases centripetal force, which increases earth's net gravity, or both 1 and 2.

Both the direct method and indirect method of increasing Earth's gravity are from multiple impacts over a long time. Changes in mass and rotational velocity are based on cumulative and multiple impacts on earth over millennia, not by one asteroid.

We know the moon has been inundated by asteroids in the past just by looking at it. The moon lacks the erosive and active dynamics that the earth possesses to cover up the scars of prior impacts. Since the moon was impacted so often, then so too was the earth inundated by asteroids in the distant past. This occurred when our solar system was young and asteroids filled the space in between planets.

There are two asteroid belts remaining, one that fills the space between Mars and Jupiter, and the Kuiper Belt that is just beyond Neptune. They are remnants that suggest what all of space was like before the gravity of each planet, star, and galaxy absorbed those asteroids and meteors into their mass. The asteroid belts could be the last vestiges of what our solar system looked like when it was young.

The amount of mass of the asteroid impacting the earth determines how much mass is added to the earth. The mass of the asteroid, multiplied by its acceleration, determines how much force hits the earth and potentially determines how much energy is taken out of the earth's rotational velocity to slow the spin down. Both methods increase gravity, one directly by adding mass and the other indirectly by slowing the rotational velocity. Both over time have the accumulated effect of increasing the earth's gravity.

How could asteroids hit the earth without obliterating all life on earth? From a Biblical perspective, there were two past events with large impacts. The first and largest occurred on the first day of creation with the creation of formless matter that began to rotate and coalesce, but there was no life yet. The image of matter coalescing is similar to a hurricane, but instead of moisture coalescing, it was matter. And instead of an empty hole at the core, it was matter. And instead of air encompassing a hurricane, it was water. The second event was a series of asteroids impacting the earth that occurred after life forms existed on earth. So how did living beings survive? Imagine a series of large asteroids hitting the earth. The impacts would devastate life on earth globally, unless there was a means of reducing the severity of the aftermath of the impacts. When those large asteroids came closer and closer to Earth, before reaching the atmosphere, they would have to pass through a canopy of salt water surrounding the earth's atmosphere (created on the second day of creation). The asteroids would cause the canopy to  succumb to Earth's gravity and come down upon the earth as rainfall (40 days and 40 nights of rain; see Gen. 7). This global rainfall would have quelled all fires and explosive effects. When asteroids hit the earth's crust, this would have fractured Pangaea, releasing 1,650 years of tectonic potential energy, and the moving tectonic plates would have squished massive amounts of water hidden in caverns under the crust of the earth and caused the water to burst out of the earth (Gen. 7:11 "the fountains of the great deep burst open."), thereby reducing the devastating effect of the asteroid impact. Thus, when the global flood covered the earth, there would have been massive amounts of soil in the floodwater that would have eventually settled down, forming the many layers of the crust.

A series of large asteroids impacting the earth would cause a massive extinction of most species near the impact zone. A bottleneck of all life or a narrowing of the family tree of life on earth would result from several large asteroid impacts. But there is a way for the severity of multiple large asteroids hitting the earth to be reduced enough for life to survive, albeit significantly reduced in its quantity and diversity. And that method is with the canopy of salt water that surrounded earth's atmosphere. This canopy would have been instrumental in quelling the devastating effect from asteroid impacts, yet the Flood would have killed all life on earth that was not protected in a floating barge called the Ark.

With the account of the Biblical Flood in Gen. 7, there is a plausible way of having large asteroids hit the earth, directly adding mass to the earth without destroying all life on earth. Plain and simple, asteroids have mass, and adding their mass to the earth's mass would increase the earth's mass, thus increasing the earth's gravity. This was probably one of the main ways God used to increase the mass of each celestial body during their coalescing days of Genesis creation. But this is not the primary means that God used to increase gravity on earth from pre-Flood to post-Flood. For there needs to be too many asteroids hitting the earth to increase its mass enough to change it 0.1%.

Let's focus on the second aspect of how asteroids can increase gravity, and that is by an indirect method. The indirect method of increasing earth's gravity is with multiple asteroid impacts slowing earth's rotational velocity and thereby increasing gravity. The cumulative effect of large impacts hitting the earth could alter the earth's spin by decreasing the earth's rotational spin velocity and subsequently decreasing the centripetal force that counteracts gravity and thereby increasing earth's gravity.

This is demonstrated when a basketball player spins a basketball on his or her finger. With an incorrect tap of a finger on the ball, the ball slows its rotational velocity, loses its centripetal force, gravity takes over, and the ball falls. It's the same concept with multiple large asteroids hitting the earth. Each large impact would take minuscule amounts of energy out of the earth's spin, until an accumulated effect occurred and the earth's rotational velocity was reduced, causing a net increase in gravity.

A large enough asteroid, or a series of large asteroids impacting the earth, would cause the earth to lose its rotational velocity. The rotational velocity produces a force called centripetal force. Centripetal force uses up some of gravity's strength and acts against gravity and thereby reduces the affect of gravity. Cosmologists concur with this concept and point to the effects of asteroid impacts altering the spin of the planet Venus. Venus spins in an opposite direction to Earth and the other planets in our solar system. Cosmologists attribute the cause of the reversal of spin to large asteroid impacts that slowed Venus' original spin to zero and then subsequent impacts that reversed the spin slightly. And this explains why Venus has a reversed spin that is very slow; one day is equal to about eight Earth months.

Does the Bible discuss asteroids impacting the earth that alters its spin velocity or the perception of time? Yes. It happened on a smaller scale in the past and it will happen on a global scale in the future. Joshua 10:12–14:

> Joshua spoke to the LORD, 'O sun, stand still'. . . So the sun stood still, and the moon stopped, until the nation avenged themselves of their enemies . . . And the sun stopped in the middle of the sky and did not hasten to go down for about a whole day. There was no day like that before it or after it, when the LORD listened to the voice of a man; for the LORD fought for Israel.

This is from the perspective of the viewer on earth. A simple explanation is that the tectonic plate that they were standing on moved parallel with the sun's normal trajectory of setting in the west. This action would have prolonged the day. The tectonic plate that they were standing on was moved from an asteroid impact, and this movement seemed to appear as the sun stood still for about a whole day for potentially ±10 hours. Just prior to Joshua recording that the sun stood still in Joshua 10:12–14, he describes God striking and killing the wicked people in verses 10–11 with:

> The LORD confounded them before Israel, and He slew them with a great slaughter at Gibeon and pursued them . . As they fled . . . the LORD threw large stones from heaven on them as far as Azekah, and they died; more died from the stones than those whom the sons of Israel killed with the sword.

This is exactly what we would expect to read if an asteroid caused a tectonic plate to slide along the path the sun moves. It indicate that the "large stones from heaven" were fragments from a large asteroid that impacted the tectonic plate they were standing on. The meteors or fragments killed more people than Joshua and his armies did by sword. This fits perfectly with asteroids and meteors hitting the earth, killing a bunch of people. And the force from the impacts caused a section of the surface crust of earth to slide and mirror the rotational spin so that the sun seemingly stood still and prolonged the day. The tectonic plate slide on top of magma until the inertia of all the matter on earth and the spin of the iron core and inner mantle caused that particular tectonic plate to resume its natural movement parallel with the earth.

Also, it happens on the Day of the Lord, when God comes to judge in the end of days, or the end times (Revelation 6:13). "The stars of the sky fell to the earth, as a fig tree casts its unripe figs when shaken by a great wind." And Isaiah 24:18–20 says, "The windows above are opened, and the foundations of the earth shake. The earth is broken asunder, the earth is split through, the earth is shaken violently. The earth reels to and fro like a drunkard and it totters like a shack." This describes a future judgment upon earth. *The windows above are opened*, indicating that at least some of the destruction brought upon the earth is from above, potentially asteroids. This adds to the idea of asteroids hitting the earth and altering the earth's rotational velocity, just like when a top has lost its spin, it starts to reel to and from and wobble *like a drunkard and it totters*. the same language is associated to lost spin and subsequently lost balance. Revelation 16:17–21 gives more details about the Isaiah 24:18–20 event:

> And there were flashes of lightning and sounds and peals of thunder; and there was a great earthquake, such as there had not been since man came to be upon the earth, so great an earthquake and mighty, the great city was split into three parts, and the cities of the nations fell . . . and every island fled away, and the mountains were not found . . . and huge, hailstones, about one hundred pounds each, came down from heaven upon men.

This seems like massive asteroids passing through the atmosphere at around 30,000 mph, causing sonic booms and peels of thunderous sounds before they impact the earth. Upon hitting the earth, they would cause massive earthquakes and split the earth asunder (potentially down the Mid-Atlantic Ridge), with some of the large asteroids splashing into the oceans and sending water up into the outer atmosphere and space. The earth's gravity would bring down some of the water as 100-pound hailstones.

Scientists have already calculated how large of an impact it would take to cause the earth's rotation to stop all together, and the Bible talks about asteroid impacts altering the earth's rotational velocity. And for this reason, it is very plausible that asteroids in the past could have impacted the earth, altering the rotational velocity and subsequently increasing the earth's net gravity.

This concept of asteroids impacting the earth and reducing the spin of the earth and thus reducing the centripetal force that tends to counteract gravity and thereby increasing gravity is destructive, and could have been a sole source of increasing gravity. But there is a greater method ahead in our discussion that is not as destructive. So we'll relegate the impact hypothesis as a contributing factor, but not as the only means of increasing gravity.

Review: Asteroids and comets struck the earth with high frequency in the past and with great size, which increased earth's gravity by increasing earth's mass and by slowing the rotational velocity of earth. This reduced centripetal force and centrifugal force and thus increased the earth's gravity. Asteroids will hit the earth again and cause the earth to wobble like a drunkard in the future. But this is not the sole option by itself for a young earth scenario to explain how gravity increased from before the Flood to after, but it may be a contributing factor.

What else could have caused a change from a weak gravity level before the Flood to a stronger gravity level after the Flood? There is another way of reducing the spin velocity of Earth and subsequently increasing gravity besides asteroid impacts. Let us focus on this rotational velocity to see if this was involved in altering gravity from prehistoric times to today.

The spin of the earth, coupled with gravity, creates centripetal force, which is center seeking. One way to describe how centripetal force tends to counter gravity is that some of gravity's energy is used up in the spin of the earth. The by-product of centripetal force that opposes gravity is centrifugal force, which is center fleeing. *Photo credit: Wikipedia.org/swing ride.*

Today, the earth's rotation velocity is approximately 1,037 mph. This rotational velocity is on an axis spinning like a top. And this is the cause of our days. It seems that spinning at 1,037 mph is a high speed, but the size of the earth considerably reduces the effect, and since we are traveling with the earth, we can't feel the velocity. But this spin velocity still effects the earth's gravity.

This velocity of earth's rotational spin creates a real force that tends to counter gravity called centripetal force (which means "center seeking") and a perceived force, called centrifugal force (which means "center fleeing"). The centripetal force is used in physics because it's calculable; however, the centrifugal force is discussed by occupants at amusement park rides because that is the force that is felt and seen. The rotational velocity of the swinging chair merry-go-round creates the apparent centrifugal force that individuals feel pushing them away from the center of the ride, and they see the effect with hair standing up on end. While the centripetal force is also felt on the swinging chair merry-go-round, it's a real force, which the chair applies to the occupant sitting in the chair.

The combination of the earth's rotation velocity and gravity results in centripetal force. The greater the spin of the earth, the more gravity is absorbed/used up, resulting in a weaker gravity level. Also, the faster the spin of the earth, the greater the centrifugal force, which is the same force generated on the swinging chair merry-go-round. The relation between centrifugal and centripetal force is the rotational velocity.

Review: Faster spin = decreased net gravity.   Slower spin = increased net gravity.

As a result of gravitational changes from centripetal force and the moon's gravity, the earth is not a sphere; it is ellipsoid (fatter at the equator than poles) in shape. This is caused by several factors, and they each deal with altering the effect of gravity. One of those factors is the rotational velocity of the earth. *Image credit: Wikipedia/org/centripetal force.*

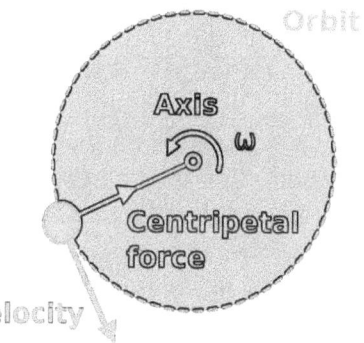

Another example of how spin can alter the strength of gravity is the weight difference from being at the equator versus being at the poles: a 200-pound man will weigh slightly less at the equator (199 lb.) and slightly more at the poles (201 lb.) because a percentage of gravity's strength is lost due to the rotational velocity of the earth. Gravity at the equator is approximately ($\approx$) 9.78 m/s/s (meters/second/second), while gravity at the poles is approximately 9.832 m/s/s. Therefore, gravity is $\approx$ 0.5% weaker at the equator than gravity at the poles. This is because of the rotational velocity of the earth. In physics, this is because some of gravity's energy is lost to rotational velocity at the equator called centripetal force. While at the poles, there is no loss of gravity's energy due to rotational velocity. Centripetal force: A force keeping an object moving on a tangential (perpendicular, like a "T" path) velocity to the earth's radius.

Mathematics demonstrates the changes in gravity with spin. Now, we know the earth's rotational velocity is slowing over time. We'll study shortly what causes the earth to slow its axis spin, but for now, let's speed up the rotation velocity of the earth as though traveling back to creation. Let's see what effect this has on gravity via calculations (the calculations are detailed in *Origins*).

For a **24-hour day:** Centripetal force at the equator is 0.034 m/s/s. Thus, Polar Gravity ~9.832 m/s/s, and Equatorial Gravity = ~9.789 m/s/s. The difference between spin (equator) and no spin (poles) today is ~a **0.5% change in gravity.**
For a **17-hour day**: The Centripetal force is 0.074 m/s/s, which **equals a 1.14% decrease in gravity.**
For a **1.5-hour day:** The Centripetal force equals the force of gravity; thus, **gravity is 100% reduced.**

This establishes that the spin of the earth does indeed affect gravity. What we are left with is the fact that spin reduces the net effect of gravity for someone standing on Earth's surface, and a 17-hour day would reduce gravity by ~1.14% from the equatorial gravity of today. Perhaps 1.14% doesn't seem like much, but it does have an effect because we are focusing on a cumulative effect, not a one-time ordeal. For example, if NASA engineers are off by 1%, then they don't land on the moon and so on.

It should be pointed out that the further back in time, the faster the earth spun. Well, we cannot go too far back in time, or else the earth will have zero net gravity and a spin velocity of one rotation in 1.5 hours, which equals zero gravity. That is not sustainable for life, because even before we get close to that velocity, the winds would be enormous, and way before that, the magnetic field would vaporize the earth's surface. But for spin, the further back in time, the faster the earth spun. According to the evolutionary model, the further back in time, the simpler the life form, with the first life occurring around 3.5 billion years ago. Well, since the rotational velocity of the earth is progressively faster the farther back we go, then we cannot go too far back in time because then the earth would be spinning too fast and gravity would be zero.

Another reason that an object is slightly lighter standing on the equator versus at either the North or South Pole is the distance of the object from the core. The earth is not a perfect sphere. It is ellipsoid (oval/oblong). Earth's radius (the distance from the core to the outer crust) at the equator is about 21,500 meters more than its polar radius. Consider the inverse square law: The greater the distance from the center of the earth, the less earth's gravity affects nearby objects. Therefore, the effects of gravity are fractionally weaker at the equator than the poles because of the difference in distance to the primary source of earth's gravity, the iron core. This is very slight and wouldn't be felt by anyone comparing the two different locations.

In our discussion of origins and life on earth, one of the contentions of this book is that dinosaurs lived in a weaker gravitational system. To get to that premise, we have to build a foundation first and then build on that foundation. The first foundation of determining what the environment was like for dinosaurs and early man is that the net effects of gravity can be weakened under certain conditions. Proving that gravity is not a constant and can be altered with several different conditions allows for the viable premise that gravity today is not of the same intensity as when dinosaurs and early mankind lived. Gravity can be altered by the spin of the earth, the radius of the earth, by the addition of asteroid mass to earth, and by the reduction of energy in the spin of the earth caused by asteroid impacts. Yet, those aren't the primary contributing factors to altering earth's gravitational force in the past before the Flood.

One of the means of altering earth's gravity is by reducing the rotational velocity of Earth. We discussed the notion that the earth's spin can be reduced by many asteroids hitting earth and reducing the spin that way. Are there other ways to reduce Earth's spin without the destructive means of asteroids? Yes, but before we explain the answer in detail, let's determine whether Earth's spin has changed over time or whether the rotational velocity of the planet has been constant from millennium to millennium. Is Earth's gravity changing over time, or is it constant? If it is changing, what is causing the increase in gravity and how?

The answer is yes; the earth's rotational velocity is slowing down over time. Yes, this increases the net effects of gravity. What is the cause of earth's rotational velocity slowing? The moon.

The moon's mass results in the moon having its own gravity that affects the earth. The moon's gravity causes our Earth to be ellipsoid and causes our ocean tides (the rise and fall of regional ocean levels), and the tides take energy out of the earth's rotational velocity (slowing down the earth's spin). Newton's first law of motion states that a body in motion will stay in motion until an outside force acts upon it. In this case, the body in motion is the earth spinning like a top, and the outside force acting upon it is the moon's gravity, which takes energy out of the spin and slows it down.

Moon's gravity → tides → ↓ Earth's rotational velocity → ↓ centripetal force → ↑ Earth's gravity.

The moon's gravity causes the ocean tides, which slow the spin of the earth by taking angular momentum (spin energy) away from the rotational velocity of the earth. What does the moon do with this energy that it takes from the earth? The energy propels the moon away from the earth. Therefore, the moon negatively affects the earth's centripetal force by slowing the earth's spin and positively affects the earth's net gravitational force by increasing gravity.

The moon is currently receding away from earth at 3.8 cm (1.5 inches) per year. Considering the inverse square law means that when the moon was closer to earth, the moon's gravity took more angular momentum out of the earth's spin velocity. Essentially, the moon slowed the earth to a greater degree when the moon was closer to the earth than today. The closer the moon was to the earth, the greater its gravitational effect was on the earth's ocean tides. Therefore, the moon had a greater effect on our tides in past millennia than today. And in past millennia, the tides were greater, which took more energy out of the earth's rotational velocity. This suggests that the earth's spin velocity is slowing at a slower rate today than in the past, which means that the further back in time we go, the faster the earth spun, and the greater the ocean tides exponentially.

This works both ways; the earth's gravitational pull affects the moon's tidal bulge as well. The moon doesn't have a body of water to physically demonstrate the effects of tidal forces. But what the moon does have are ancient lava beds from when the moon was active. Those ancient lava beds are the dark spots on the moon, and they represent a time when the moon was active and hot because it was closer to Earth. When the moon was active and hotter as a result of the earth's gravitational influence on the moon causing friction, molten rock flowed onto the surface of the moon and formed the dark spots we see from Earth. The moon didn't spin fast like the earth spins to cause the tidal frictions, but the moon orbits the earth in an elliptical pattern. This means that the moon does not orbit the earth in a perfect circle. We can visualize this elliptical pattern by observing the moon appearing larger in the sky at its perigee (proximal) and smaller in the sky at its apogee (distal). In the past, when the moon's elliptical perigee and apogee were both closer to Earth, it caused internal tidal friction within the moon and generated enough heat for lava to flow out onto the surface. The earth's influence on the moon's internal tidal friction and subsequent heat was increasingly stronger the further back in time, as the moon was closer to the earth. But now that the distance from the moon and the earth has increased, the earth's influence that once caused tidal friction and subsequent lava to flow on the moon has ceased, and the moon is minimally active.

Since we know the moon is using the angular momentum it takes from earth's rotational velocity to move away from the earth, then the earth's rotational velocity was greater in the past. When the moon was closer to the earth, the moon slowed the earth's rotational velocity to a greater extent because the moon's gravity had a greater effect on the earth's ocean tides. And the greater the ocean tides in past millennia, the more angular momentum the moon took out of the spin velocity of the earth. This proves that the earth's net gravity was progressively weaker, the further in the past.

Review: The moon recedes away from earth at 3.8 cm per year. The moon was closer to the earth in the past. The earth's spin is slowed by the moon's gravity. Since the moon is getting further away from earth, its gravity takes less angular momentum out of the earth's spin today than it did say 5,000 years ago. Therefore, the earth is slowing at a slower rate today than in the past, which also means that earth spun exponentially faster in the past than today. This means that the centripetal force that countered against gravity was greater in the past. Therefore, gravity is indeed exponentially weaker the further back in time we go.

Why couldn't the earth-moon relationship be billions of years old? Why not say that the effect the moon has today is weaker than the effect the moon had on earth billions of years ago? There are several reasons, but for the sake of this discussion, we'll address one of them. Since the moon is moving away from the earth at 3.8 cm (1.5 inches) per year, then as we go backward in time, the moon gets closer to the earth. There is a point were the moon cannot be any closer to the earth, or the gravitational forces of the moon and earth would cause disintegration of one or both bodies. This distance is called the Roche limit. But before the Roche limit is reached, tidal forces would be too large for life to be sustainable.

The rate at which the moon is receding away from the earth is gradually slowing. Physicist Donald DeYong explains, "One cannot extrapolate the present 4 cm/year separation rate back into history. It has the value today, but was more rapid in the past because of tidal effects. In fact, the separation rate . . . was perhaps 20 meters per year long ago, and the average is 1.2 meters per year."

Below is an idea of how the moon's recession away from earth could have looked. These numbers are only to give a sample of how the moon's recession is not linear but parabolic.

Century	Change in a Day's duration/100yrs.	Rate of Moon's recession	Gravity	Length of Day
21st century	slowed 2 msec./100yrs.	3.8 cm/yr.	9.8m/s/s	24 hours
15th century	slowed 5 msec.100yrs.	7.5 cm/yr.		
10th century	slowed 25 msec./100yrs.	15 cm/yr.		
1 BC	slowed 500 msec./100 yrs.	85 cm/1 yr.		
1,000 BC	slowed 12 sec./100yrs.	400 cm/yr.		
2,000 BC	slowed 1 min./100yrs	1 meter/yr.		~23 hours
3,000 BC	slowed 7 min./100yrs.	5 meter/yr.		~22 hours
4,000 BC	slowed 30 min./100yrs.	20 m/yr..	~9.6m/s/s	~17 hours

The above also suggests how the length of a day around the time of Adam and Eve could have been about 17 hours long. It has been increasing ever since. This means that the earth would be spinning fast enough to weaken gravity by 1.1% for Adam and Eve. This by itself is not enough to appreciably alter life on earth. Yet, this adds to the cumulative effect on net gravity and the subsequent effect of life on Earth. With the moon closer to the earth, the moon's gravitational force would directly decrease the earth's gravitational force by pulling against it. This is different than the discussion above; this is simply the moon's gravity pulling in an opposite direction to earth's gravity, thereby decreasing earth's gravity directly via subtraction.

Gravity is not even uniformly constant on earth; it's not 9.8m/s/s all around the earth. There are regions on earth where gravity is greater due to changes in density beneath the crust of the earth, changes in elevation, and centripetal force. So since gravity is not even constant on earth, why would anyone think it's been constant throughout the millennia?

With the moon's gravity and the rotational velocity of earth being able to alter the physical geography of the earth from a sphere to an ellipsoid (oblong), this is good evidence that they should also affect life on earth as well. This begs the question, "When the moon was closer to the earth and had a greater pull on the earth's shape and the earth was spinning faster, was the earth's size larger as well?" Yes, the closer the moon's gravitational pull is to the earth, the greater the earth becomes oblong.

Scientist Donald Hamilton suggests a 10% increase in the earth's equatorial bulge from a faster rotational velocity upon its axis. The fast spinning star Vega makes a full rotation about its axis once every 12.5 hours, which causes it to assume a 23% elliptical bulge at its equator, so a 10% greater bulge for the faster spinning Earth is not out of the question. Essentially, this suggests that the faster the earth spun in past millennia, the fatter the earth was, perhaps with a 10% greater equatorial circumference.

Since the earth's rotational velocity is slowing down and the rate at which the earth is slowing down is decreasing, this proves that earth had a higher rotational velocity in the distant past, which is strong evidence that the earth had a larger equatorial bulge in past millennia than today. Why is this important? The inverse square law suggests that the farther an object is from the source of gravity (the central iron core), the less gravity affects the space and time of an object. Therefore, an increased size of the earth means that the inhabitants are farther away from the iron core center of the earth, which is the primary source of its density and gravity. Below is a chart of earth's gravitational pull at different altitudes:

Altitudes (km)	Gravity (m/s/s)
0 km surface level	9.81
31 km = 20 miles	9.74* 1% weaker gravity
62.5 km	9.67
125 km = 77.5 miles	9.50* 3.2% weaker

You can see that the farther away from the source of gravity, the weaker gravity becomes, and the primary source of gravity is the iron core. We have already established that the earth is larger at the equator than the poles because of the moon and spin and that the earth was larger in the distant past than today at the equator because of the moon's gravity and the faster spin of earth. So if the earth's surface were 20 miles further away from the iron core, then gravity would be an additional ~1%± weaker at the equator. Given that our earth currently has a radius at the equator of 6,371 km, an increase of 31 km more to the radius is a 0.5% increase in the radius of earth. This is certainly not out of the realm of plausible. If the earth had an equatorial bulge of 77 miles (2% equatorial bulge), then gravity would be approximately 3.2% weaker. Even though this is a plausible explanation of how gravity was weaker in the past, it's still not the primary reason for a reduced gravity that allowed dinosaurs to grow so tall. It's just a contributory factor, but all these contributory factors start to add up.

Review: Since the moon was closer and pulled on the surface of earth, the circumference of the earth would have been larger when the moon was closer. And since the earth spun faster in past millennia, the faster spin resulted in greater centripetal force, which resulted in earth being a slightly larger planet. Therefore, gravity's effect was weaker on earth's inhabitants in past millennia because they were further away from the iron core. Just a 0.5% increase in the radius of earth amounts to an additional ~1% reduction in net gravity.

Since the earth's rotational velocity is fractionally slowing each year, that means the centripetal force countering gravity is fractionally getting weaker each year as well. And since the earth's mass is continually being added to by space debris each year, then the earth's mass and subsequent gravity are fractionally getting higher each millennium. And since the moon is moving away from earth each year, and the moon's gravity counters earth's gravity, then the earth's gravity is fractionally getting higher each millennium. Therefore, the net effect is that gravity has been increasing each millennium. But these changes are fractional and may only amount to a 1%–3% change in earth's gravity.

With gravity gradually increasing, this increases the internal heat of the earth, which explains the increase in earthquakes and volcanic activity today. This parallels how God describes His judgment in the day of the Lord, with earthquakes (Matt. 24:7) and fire (2 Peter 3:7).

The earth's length of day is getting longer each century by approximately 2 milliseconds per 100 years. The rate of lengthening of the day is not constant. The rate has been getting gradually shorter and shorter with each century. In other words, if the length of the day is getting longer by 2 milliseconds today, it's incorrect to think that in 5,000 BC, the rate was the same 2 milliseconds. This is wrong for one main reason—the moon. The moon is one culprit (asteroid impacts are another) for the earth slowing, and the further away the moon gets from earth, the less effective it becomes at reducing the earth's rotational velocity. Below is a graph of the relationships between the moon's distance from the earth, the length of a day, and the earth's rotational velocity based on distance and time. The dashed (· · — ·) line represents the incorrect views for each.

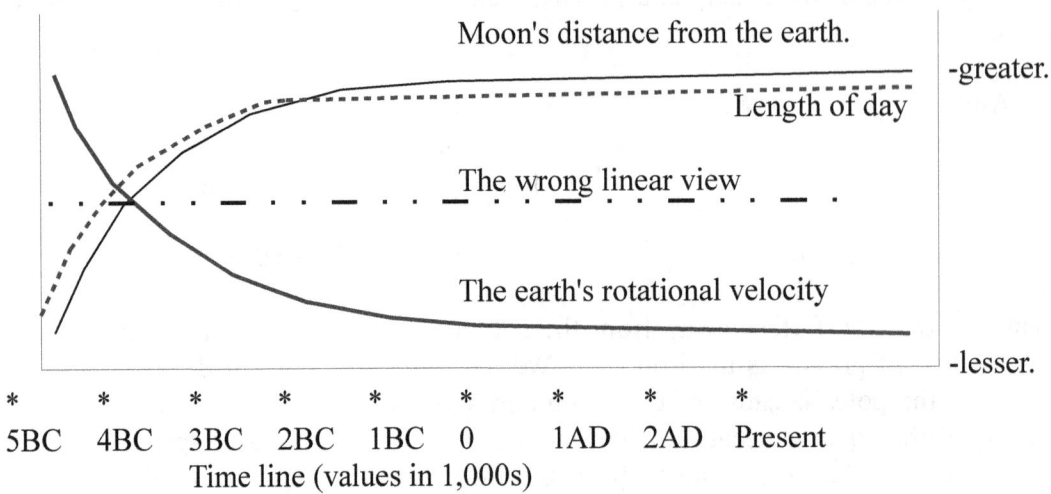

Review: Whether it was by asteroids impacting the earth and adding mass or reducing spin, or whether it was the moon's gravitational force altering the space and time near the earth and changing the earth's gravity or the moon's gravity taking energy out of the spin of the earth, the bottom line is that earth's gravity has changed, and all of the above are cumulative contributors.

Now that we have established that the earth's gravity can change and has changed from past millennia, it's time for another piece of the gravitational puzzle. This is the final bit of evidence that the net gravity was weaker some 5,000 years ago. This final bit of evidence that changed gravity on earth is buoyancy—via the canopy of salt water that hovered spherically around the atmosphere and caused the global flood of Gen. 7.

How would a canopy of salt water change gravity on earth? It is that most of the ocean water—before being formed from the torrential rain—would be hovering above the atmosphere and thereby would increase atmospheric pressure, weight, and density of air.

How does increased atmospheric pressure reduce gravity? Let's use water as an example to set the stage because the concept of buoyancy, which atmospheric pressure provides, is similar to the buoyancy that a pool of water provides, just at a significantly reduced level. This is based on Archimedes' principle of buoyancy force. When a body is submerged in water, either wholly or partially, it is buoyed up by a force equal to the weight of the liquid displaced by the body. When you get into water, a certain amount of volume of water is moved out of the way to make room for the volume of your body. The mass of that amount of water displaced will create a force lifting you up. If your body weighs less than the weight of the volume of water displaced, then you'll float. If your body weighs more than the weight of the volume of water displaced, then you'll sink.

Therefore, if you ever wondered why one floats in water, it is because fat cells are less dense and weigh less than an equal volume of water displaced. The buoyancy force is greater because the volume of water displaced is denser and weighs more than a body filled with large amounts of fatty tissue that is less dense and weighs less. Subsequently, this buoyancy force is then greater than the force of gravity needed to pull an individual down to the bottom of the pool, so the individual floats.

Conversely, top athletes sink because muscle mass is denser and weighs more than the equal volume of water they are displacing. Therefore, the buoyancy force is weaker because the volume of water displaced weighs less than the muscular body, and thus buoyancy force is then weaker than gravity; thus, they sink.

Here's the formula: Buoyancy = weight of volume of liquid displaced − body weight.

If the sum is < 1, the body sinks. If the sum is > 1, the body floats. If the sum = 1, the body doesn't sink or float, but remains suspended.

Can the amount of buoyancy force change? Yes, for example, a body of water with a very high saltwater content will have a greater buoyancy force than fresh water. Why? The salt water has more particulates per cubic volume of water, which means the salt water is denser. For example, in the Dead Sea in Israel, the water is so dense from the high salt concentration that one floats almost at mid-chest level in the water. That same person may sink in fresh water. Therefore, we see that buoyancy can change with changes in the density of water. Since Archimedes' principle equally applies to whatever medium an object is immersed into, including air, the buoyancy of our atmosphere (though less dense than water) can still change the amount of buoyancy force it applies against gravity depending on the density of air.

What is an example of how changes in atmosphere pressure affect buoyancy? Fishing. For those that want to know the best time to fish, it's when there is high pressure above the water. Why? High pressure or high barometric pressure (synonyms) means that there is more atmospheric weight pressing down on the water. This means that the weight of the water that the volume of fish displaces will weigh more. This change in buoyancy force causes the fish to float. Hence, the fish become more active to prevent them from rising and floating to the surface.

Another example is when someone has a painful joint and a storm is approaching. They are the first ones to tell everyone that their joints ache. Why? When a storm is approaching, the atmospheric pressure drops, and therefore there is less pressure. With less pressure, swollen tissue, scar tissue, and bone spurs expand and physically press on the nerves, sending pain signals to the brain. It's like their injured joint is a weather detector.

A really good example of the similarities of the buoyancy of air and water is the comparison of a jellyfish and a balloon. As starfish look up through an ocean of water to see jellyfish floating in the water, humans look up through an ocean of air to see balloons floating in the air. Both the jellyfish and the balloon are buoyed upward by a force equal to the weight of volume of fluid displaced by their volume.

The effect of a heavier atmosphere, creating higher atmospheric pressure, has the same application of Archimedes' principle of a larger buoyancy force, except that it is air being displaced by your body, rather than water being displaced. And just like water, the denser the air, the more buoyancy force is applied. The more air matter per cubic volume, the denser the air and the greater the atmospheric pressure, resulting in more matter being displaced by the body being in the atmosphere, which is a greater buoyancy force. Now, this is all fine and dandy up to a point. There comes a point where body tissue can't handle too much pressure, or it starts to collapse on itself. Much like a submarine that sinks to the bottom of the ocean, when the pressure near the bottom of the ocean is too great, the submarine collapses like a tin can under a stomping shoe. Conversely, if there is not enough pressure, then our body tissue can't hold itself together (we boil, not from heat, but from a change in pressure) and we explode. Just like when a balloon floats too high in the atmosphere and the pressure is too low, the balloon will expand and expand until the integrity of the walls explode. You may have seen the science fiction movie in which an astronaut loses pressure from being in space and explodes. Variations in pressure and density can have a profound effect.

<u>Vacuum of space</u> → Zero pressure: humans explode.
<u>High altitude</u> → Very little pressure: balloons burst.
<u>Surface of the earth</u> → 1 Atmosphere of pressure: today's normal.
→ Somewhere in here is the optimal pressure: Adam and Eve's normal.
<u>Fresh water</u> → Medium pressure: high buoyancy.

Salt water   → Higher pressure: higher buoyancy.
Ocean floor   → Extreme pressure: submarines are crushed.

You can see that with different amounts of pressures, there are varying results. At zero pressure, the body will boil on the inside and may explode. Too little pressure, and a balloon explodes; too much pressure, and a submarine is crushed. By looking at slight variations of pressure, there are beneficial results to buoyancy. Somewhere between one unit of atmospheric pressure (at the surface of the earth) and fresh water is a sweet spot of buoyancy force that existed in our atmosphere before the Flood and subsequently resulted in a reduction of net gravity.

Why is air a factor in terms of buoyancy, even though we can't see it as we can with water? Air has mass just like water has mass, but air has a lesser amount per the same cubic volume. Archimedes' principle states, "Any body completely or partially submerged in a fluid is buoyed up by a force equal to the weight of the fluid displaced by the body." And gases apply to this principle. To test whether air has mass, just roll down your window while driving in a car at 55 mph.

What causes buoyancy in air? The pressure of the atmosphere is pulled down by gravity, and when the air reaches the surface of the earth, those air particles literally push up on the rest of the atmosphere and hold up 400 miles of air mass. This pushing up has a force, and that force is called buoyancy. Therefore, when a body is standing on the surface of the earth, that body is displacing a certain amount of volume of air, and that volume of air has mass and weight that creates a buoyancy force that lifts up.

Remember: Buoyancy force = the weight of object – the weight of air displaced.

The more volume of air per cubic volume displaced by an object, the more weight of air displaced and the greater buoyancy effect pushing upward in the opposite direction of gravity. Therefore, though gravity has not changed, its net force is subsequently reduced as a result of buoyancy force.

The heavier the atmosphere, then the more weight that the particles of air have to hold up near the surface of the earth, then the more pressure they exert on people on the surface of the earth and, subsequently, the more buoyancy force applied per volume of cubic air displaced. With greater buoyancy, the force of net gravity is weaker. All we need to do is determine if our atmosphere at one time had greater weight pushing down on earth; if so, then the air particles nearer the surface would have to push up equally as hard, and this would increase the buoyancy force, which would reduce net gravity. This is where the rubber meets the road. Before the flood of Gen. 7, there was a canopy of salt water that hovered around the atmosphere. This canopy of salt water had a massive amount of weight.

How do we know there water surrounding the atmosphere? There are five reasons:

No. 1: The Scriptures (Gen. 1:6–10):

> God said, "Let there be an expanse in the midst of the waters, and let it separate the waters from the waters [water above and water below the atmosphere]." God made the expanse, and separated the waters which were below the expanse from the waters which were above the expanse; and it was so. God called the expanse heaven. And there was evening and there was morning, a second day. Then God said, "Let the waters below the heavens be gathered into one place, and let the dry land appear [Pangaea]"; and it was so, God called the dry land earth, and the gathering of the waters He called seas.

When God created the atmosphere, He did it by separating water from water to create an expanse/firmament/heaven. The waters below the heavens became the seas. The waters above the expanse/heaven/firmament/atmosphere were still there above the atmosphere, waiting for a purpose, which came later in the Gen. 7 global flood.

No. 2: Deduction: The source for enough water to flood the earth in Gen. 7. Since Gen. 2:5–6 indicates that there was no rain until the Flood, then the water came from beyond the atmosphere, and the canopy of water is a plausible explanation. Rain falling over the entire earth for 40 days and 40 nights required a lot of water, and that massive amount of water had to come from somewhere. The canopy is a perfect explanation and is supported by Scripture. There is no doubt that a global flood occurred. Every mountaintop has sea shells, and almost every culture of mankind has some historical record of a flood. However, if you are an evolutionist, you are forbidden from accepting a global flood scenario because a global flood explains all fossils, coal deposits, oil deposits, salt deposits, layers of sediment on the crust of earth, high oxygen concentrations in the past, reduced gravity in the past, glacial ages, and finally a young earth. And this is unacceptable because billions of years are needed for evolution.

No. 3: Oxygen concentration: Before the Flood, oxygen concentration was 50% higher than after the Flood. This is discerned from glacier core samples and air bubbles trapped in amber. For 50% higher oxygen concentrations to exist, the earth would need to have more exposed and usable land for vegetation to grow to be able to produce that much oxygen. This means there were no oceans until the Flood, which means a portion of the ocean water was stored in the canopy surrounding the atmosphere.

No. 4: Nearly every celestial body in space has remnants of water markings or frozen water., indicating that when the universe was young, water was everywhere.

No. 5: Four of the planets in our solar system (50%) still have water hovering above their atmosphere. However, because of centripetal force, their water exists as rings—most notable is Saturn's rings.

The salt water came down upon the earth in 40 days and 40 nights of rainfall. This means that before the Flood, this canopy of salt water would have hovered above the atmosphere and would have added a great deal of atmospheric pressure, weight, and density, increasing the buoyancy force and subsequently reducing net gravity.

You may be surprised to hear that our atmosphere today weighs about 5.5 quadrillion tons (one ton = 2,000 lb.). Isn't that bizarre? This is nothing new for the Bible, as it says in Job 28:25, "He imparted weight to the wind," which was written some 3,500 years ago; it never ceases to amaze me how the Bible is divinely in harmony with true science and how seemingly wandering nomads could write in perfect harmony with what modern mankind has only recently discovered. Back to the discussion: later on in the book you'll learn that prior to the Flood, there were no polar ice caps, and all the polar ice caps were a result of the Gen. 7 global flood. Therefore, we can take the entire weight of the polar ice caps, 32 quadrillion tons, and add that to the atmosphere's weight to help determine what the buoyancy force was before the Flood. Thus, putting the weight so far of the atmosphere plus the polar ice caps at 37.5 quadrillion tons, or 3.75×10^{16} tons. The additional weight added to the atmosphere by the polar caps increases the buoyancy force by a factor of 6X. But there is still more weight to add to the canopy of salt water hovering above the atmosphere because the rainfall didn't just occur on the polar regions, but globally. Therefore, we must consider a portion of the oceans as well. It's possible that a majority of our vast oceans were formed as a result of the water coming down onto earth for 40 days and 40 nights. The oceans are estimated at 1.5×10^{18} tons. If we figure conservatively that only one-fourth of the ocean water came from the Genesis flood, then that would add 37.5×10^{16} tons of ocean water, plus 3.2×10^{16} tons of polar ice caps to the weight of the atmosphere (0.55×10^{16} tons), totaling 41.25×10^{16} tons of atmospheric weight. That is 75 times greater than our current atmosphere's weight today.

Therefore, the estimated buoyancy force before the Flood of Gen. 7 and the environmental condition of early man and dinosaurs before the Global flood would have been 75X greater than today based on only 25% of the ocean water stored up above the atmosphere. This equals a buoyancy force great enough to reduce the net effect of gravity by almost 10%.

Weight of the atmosphere:	5.5×10^{15} tons
Weight of the polar ice caps:	3.2×10^{16} tons
Weight of one-forth the oceans:	$\underline{37.5 \times 10^{16} \text{ tons}}$
Total:	41.25×10^{16} tons

This does not include the vast salt mines that were formed from the Flood, which would increase the weight pushing down on the atmosphere and increase the weight of the atmosphere and thus increase the buoyancy force, which decreases net gravity.

A cubic meter of air has a mass of 1.2 kg (12 newtons) today.
Air (today) has a density of 1.29 kg/m^3.
Fresh water has a density of 1,000 kg/m^3.
The ocean has a density of 1,027 kg/m^3 (2.7% heavier than fresh water).
Seawater has a density of 1,035 kg/m^3 (3.5% heavier than fresh water).
The average human has a density of 1,000 kg/m^3, and an average person weighs 70 g, so
    the average volume of the human body = m/M = 70 kg/1,000 kg/m^3 = 0.07 m^3.

For the average human being in **ocean water, net gravity is 102.7% reduced** by ocean water.
For the average human being in **fresh water, net gravity is 100% reduced** by fresh water.
For the average human being in **air (today), net gravity is 0.12% reduced** by air.
For the average human being in **air (pre-Flood based on 25% of ocean water stored in the canopy), net gravity is 9.8% reduced** by air pre-Flood. This is based on 25% of ocean from the canopy of water, 25% of ocean from deep caverns below the crust, and 50% of ocean already in existence as seas.

Now you can see how changes in the density of the volume of a fluid that a human body is submerged in determines the buoyancy force. If the canopy of salt water included only 25% of the ocean water, then this increased air density by a factor of 75 and reduced net gravity by ~9.8%.

Suppose a book suspended by a string weighs 10 newtons in a vacuum, with gravity acting on the book. Then imagine lowering the book into water, and in the volume of water, the book displaces 4 newtons of water. The net force exerted on the string holding up the book is represented by 10 newtons – 4 newtons buoyancy force = 6 newtons net. See how gravity was reduced by buoyancy force. This changed the net affect of gravity, and gravity was reduced by 40% in this fictitious scenario.

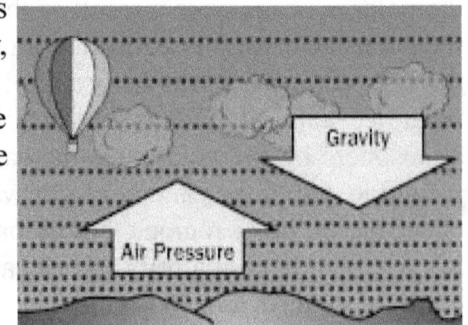

If the amount of water that rained upon the earth from the canopy in 40 days and 40 nights of global rain was 50% of the oceans, then:

The weight of the atmosphere:	0.55×10^{16} tons
The weight of the polar ice caps:	3.2×10^{16} tons
The weight of half the oceans:	$\underline{75 \times 10^{16} \text{ tons}}$
Total:	78.75×10^{16} tons

Then 78.75 x 10^{16} tons divided by 0.55 x 10^{16} tons (current atmospheric weight) = 143 (factor of change). This means that the atmosphere before the Flood weighed 143X greater than it does today. Therefore, the density of the atmosphere before the Flood was 143 times greater than today's value, with 50% of the ocean water stored up in the canopy of salt water. The formula to estimate the buoyancy force of air prior to the Flood utilizing 50% of the ocean water in the canopy of salt water is:

For the average human being in air (today), net gravity is 0.12% reduced by air.
For the average human being in **air (pre-Flood based on 50% of ocean water stored in the canopy), net gravity is ~18.1% reduced** by air pre-Flood. *Image credit: How Hot Air Balloons Work, by Tom Harris, www.science.howstuffworks.com.*

The bottom line question is, "How much water was stored up above the atmosphere as the canopy?" The answer lies in the ocean. How much of the ocean water was formed from 40 days and 40 nights of global rainfall? No one knows. But it has to be that a significant percentage of the oceans were formed from the rainfall coming down upon the earth in the Gen. 7 global flood because the air bubble samples in the glacial core samples indicate that oxygen was 50% higher before the Flood. The only way to increase oxygen is to increase the quantity of vegetation. The best way to increase vegetation is to reduce the size of the oceans that limit vegetation growth.

<u>If 25% of the ocean was stored up in the canopy</u> before the Gen. 7 Flood, <u>then gravity was reduced by 9.8% from buoyancy force</u>.

<u>If 50% of the ocean was stored up in the canopy</u> before the Gen. 7 Flood, <u>then gravity was reduced by 18.1%</u>. And this doesn't factor the weight of all the salt mines, a faster spinning earth, closer moon, larger earth radius, 50% higher concentration of oxygen before the Flood, asteroid impacts, mass being added to the earth, and so on.

We know God placed water under the surface of the earth because of Psalm 136:6: "To Him that stretched out the earth above the waters" and Gen. 7:11: "The fountains of the deep burst open." It is a safe bet that these fountains were water, and they burst out of the crust at the fault lines and volcanoes. Since most volcanoes erupt massive amounts of water as well in the form of steam and the most violent eruptions come from water contacting the hot magma and resulting in massive accelerated explosions, then quite possibly, the "fountains of the deep" that burst open could very well have included volcanoes around the globe, especially since the discovery that there are still ample amounts of water in the mantle.

Even though there would be an increase in the weight of the atmosphere, the pressure would still be far below the pressure of being in water and would be far below the density of water. Anyone who accepts that there was a flood has to put some percentage of the ocean formation from the rain that fell down upon the earth up into the canopy of salt water. Whether it's 25% or 50%, the force of gravity was reduced by the force of buoyancy by either 10%, 18%, or some percentage near or between them. Thus, a plausible explanation of how the net force of gravity was weaker before the Flood some 4,400 years ago has been established. Though the force of gravity resulting from the increased mass of the earth remained relatively stable from the time before the Flood to the time after the Flood, one can see that the force of gravity is determined by many factors, and those factors have changed over time and have altered the net strength of gravity from the origins of life on earth before the Flood to today's current intensity. The bottom line is that the net force of gravity was weaker in the past. Even if only 50% of the ocean water was stored up above the atmosphere, thereby increasing the weight of the atmosphere, it would still only result in 18% of the density of fresh water. Therefore, life forms wouldn't be crushed by the weight and pressure of the atmosphere, but would be more buoyant with a greater density, weight, and pressure of air.

Review: Buoyancy force is a real force that is felt, and the effects are observable, measurable, and calculable. It passes the scientific method of being observable and testable. Archimedes' principle of buoyancy force applies for both water and air. The denser the air, the greater the buoyancy force. The atmosphere is made denser by greater weight above it, which squishes the air molecules below closer together. Thus, before the Flood, the water being stored up above the atmosphere in the canopy of salt water created greater atmospheric pressure, weight, and air density, and the buoyancy force of air before the Flood was great enough to reduce the net effect of gravity by 10%–18%, depending on how much salt water was stored in the canopy.

In physics, there are laws pertaining to pressure, temperature, and volume for gases. They are Boyle's law, Charles law, Gay-Lussac's law, and the combined gas law:

Boyle's law: If temperature is constant, then the Pressure exerted by a gas varies inversely to its Volume, which means that: $P_1V_1 = P_2V_2$
* If the pressure increases, the volume decreases.
* If the pressure decreases, the volume increases.

Charles Law: If pressure is held constant, then the Volume of a gas is directly proportional to its absolute Temperature, which means: $V_1/T_1 = V_2/T_2$
* If volume increases, the temperature increases.
* If volume decreases, the temperature decreases.

Gay-Lussac's law of pressure and temperature: $P_1/T_1 = P_2/T_2$
* If volume is held constant, the pressure of a gas is directly proportional to the temperature.
* If the pressure increases, temperature increases
* If the pressure decreases, temperature decreases.

Combined gas law: $P_1 \times V_1/T_1 = P_2 \times V_2/T_2$
* It combines all gas laws into a simple expression.

How this applies to our discussion of the origin of life and gravity is that at the time before the global flood of Gen. 7, the atmosphere was denser because there was a canopy of water above the atmosphere that compressed it, and the oxygen concentration was 50% higher than today's level because of a greater quantity of vegetation. The temperature of the atmosphere would have been kept relatively constant because of the canopy and atmosphere, creating ambient global temperatures, and the temperatures would have been around 70°–80°F, which is warmer than today's average (14°C or 57°F). With the salt water pressing down on the atmosphere to increase the pressure, the size of the atmosphere would have been reduced as a result of weight, and the density and pressure of the atmosphere would have increased. Thus, the buoyancy force would increase.

However, the increased temperature of the atmosphere would have increased its volume and thus somewhat counter the compressing of the atmosphere from the weight of the canopy—as discussed with the Boyle's Law. With an increased temperature before the Flood, there was increased atmospheric pressure. This explains why some dinosaurs that had nostrils too small for their size still thrived because there was a greater tidal volume of air that passed through their lungs with each breath. And since there was increased oxygen concentration before the Flood, then each breath brought in more oxygen, which caused all life dependent on O2 to thrive.

The bottom line with the laws applied is that the atmosphere would have reached a natural homeostasis of equilibrium. There was more weight above the atmosphere causing increased pressure that resulted in a force that wanted to squish the atmosphere, but there was an increased temperature that caused increased atmospheric pressure that pushed back. Although the volume of the atmosphere may have been smaller because of the increased weight and subsequent pressure applied to the atmosphere, there was an equilibrium that was reached that caused life to thrive. And as someone is buoyant on an ocean of water, the canopy was also buoyant on the ocean of air and increased the buoyancy for everything on the surface of the earth.

Review: Increased atmospheric pressure results in a more condensed atmosphere. Increased temperature pushes back to prevent too much squishing. The volume of the atmosphere prior to the Flood would have been reduced to handle the weight, and all of this resulted in a greater buoyancy force, which reduced net gravity.

How do we discern that global temperatures were higher before the Flood than today? Here is a list of the evidence: (a) Adam and Eve were naked all day, 7 days a week, and 365 days per year (Gen. 3:7). (b) There were no polar ice caps because the canopy of salt water kept uniform global temperatures. (c) There was no wind before the Flood (Gen. 8:1); therefore, there were no deserts and no polar ice caps. Changes in temperature are required to generate wind. (d) There was no rain before the Flood (Gen. 2:6), as high pressure prevents cloud formations, thus evaporation bypassed the formation of clouds and rain and skipped right to a mist (dew) that watered the ground; thus, the hydrological cycle was altered (Gen. 2:6). (e) There were no rainbows before the Flood (Gen. 9:13–14), as a hydrological cycle is required to make rainbows. And (f) there were no clouds before the Flood, as the first mention of clouds in the Bible is in Gen. 9:13. Not even the record of the Flood during Noah's time mentions clouds.

Review: The weight of the canopy, increased the pressure, density, and temperatures of the atmosphere, and this increased the buoyancy effect and weakened net gravity.

We know that oxygen concentration in the atmosphere was 50% higher before the Flood than today's value (glacier core samples and amber air bubble samples). Today, oxygen concentration makes up roughly 21% of our atmosphere, and nitrogen makes up 78% (< 1% comprised of other elements).

However, before the Flood, oxygen made up 31% of our atmosphere, and nitrogen made up 68% (< 1% other elements). What's noteworthy about this is that oxygen weighs slightly more than nitrogen and is denser, so this would have added to the atmospheric pressure and density of the atmosphere. Oxygen weighs 14.3% more than nitrogen (oxygen molecular weight = 32 g/mole; nitrogen molecular weight = 28 g/mole) and is 2.5% denser than nitrogen.

Considering the above values, this would have increased the weight of the atmosphere, increasing the buoyancy effect of the pre-Flood atmosphere and subsequently reducing net gravity by 0.14%. This is negligible, but it again shows that changes in the environment can have an affect on the net effectiveness of gravity.

Review: Before the global Flood, increased oxygen concentration increased atmosphere mass, weight, pressure, and density, which increased the buoyancy effect and decreased net gravity, although it was by trace amounts. In addition, the warmer global temperatures before the Flood resulted in greater atmospheric pressure and increased the buoyancy force, which reduced the net gravitational force. The increased atmospheric pressure would have also increased the tidal volume of oxygen entering the lungs with each breath, which would have caused life forms to thrive and made life easier to sustain and would have resulted in human beings who lived 900+ years, as recorded in Genesis.

There is another minor element of the evidence that needs to be added that would have affected the weight of the atmosphere when a portion of the oceans was hovering above the atmosphere as the canopy. There are massive salt mines on the planet. All of these salt mines were formed during two catastrophic events, either during the formation of the hot earth that was surrounded by water by the process of evaporation that formed the deeper salt basins or when the global flood of Gen. 7 occurred and freezing temperatures froze fresh water out of salty solution that formed the shallower salt basins. Salt water is 3.5% heavier than fresh water. When the heavy rains came upon the earth in the Genesis flood, that highly salty rainwater saturated the waters below on the earth. After the 40 days and nights of rain, then the waters prevailed for 150 days, this caused temperatures to plummet and fresh water began to freeze out of the salty flood waters. This process caused the salt water near the fresh frozen ice to become saturated with salt. When the water reached a saturation point, which meant that no more salt could be diluted in the water, then as more fresh water was frozen for the glacial age, the extra salt left behind in the salt water could not dilute into solution and would sink to the bottom in solid form. This process formed vast salt mines.

What does salt have to do with buoyancy and gravity? The salt mines formed because of the Flood would have added to the density and weight of the canopy hovering around the atmosphere prior to the Flood, which means that more pressure and weight resulted in a greater buoyancy force and reduced net gravity. Since seawater is 3.5% denser than freshwater, then we may conclude that the buoyancy force from a seawater canopy was higher than the buoyancy force from a freshwater canopy.

We have discussed how the spin of the earth creates centripetal and centrifugal forces, which tend to counteract gravity, resulting in weaker net gravity. We have also noted that the moon takes angular momentum out of the spin of the earth, slowing the earth's spin and gradually increasing the earth's net gravity. And we have suggested that the moon was closer to the earth, having a greater effect on reducing the earth's net gravity and that the moon, while it was closer to the earth, slowed the spin of the earth to a greater extent than it slows the spin of the earth today. Moreover, we have proposed that the length of one day about 6,000 years ago was approximately 17 hours long, and this created weaker net gravity because the earth spun faster, creating a shorter day. We also discussed the notion that space debris adds to the earth's mass and increases gravity and that asteroid impacts affect gravity by adding mass and decreasing spin. Additionally, we have declared that a faster spinning Earth would have meant a larger bulge at the equator than today, which would have affected the net strength of gravity as one was further away from the iron core. And lastly, we have described how the buoyancy force from the canopy of salt water hovering above the atmosphere also reduced net gravity.

The exact percentage of how much each of those factors would have affected the intensity of gravity can only be estimated:

Conditional change:	Net reduction of Gravity:
Faster spin leading to 17 hours for one rotation	1%
Closer moon	1%
Larger Earth radius	1/2%
Asteroid impacts	1/10%
Space debris added to mass	1/10%
Increased O2 in the atmosphere	1/100%
Buoyancy force from the canopy of salt water	11–19%
Total reduction of gravity before the Flood	~13.5%–~21.5%

All of the listed items above could by themselves affect the strength of gravity. If the spin of the earth was fast enough so that a day was 1.5 hours long, then net gravity would be neutralized to 0.0 m/s^2 from centripetal and centrifugal forces. If the mass of the earth stayed the same, but its radius increased from a faster spin, then gravity would be proportionately reduced by the inverse square law, which suggests that the farther away an object is from the source of gravity, the weaker gravity becomes. But the primary reduction of gravity's net intensity came from the buoyancy force generated from the canopy of salt water that hovered above the atmosphere. Is it even possible to have water hovering above a planet similar to the canopy? Consider that Saturn's rings have ~26 million times more water than Earth. That is one example of water around a planet, albeit in the form of ice rings. Therefore, it is plausible to have had a canopy hovering above the earth.

CHAPTER SUMMARY: This chapter discussed the idea that the gravitational force can be reduced or increased depending on the environmental conditions. Also, it propounds that before the Flood of Gen. 7, net gravity was 13.5% to 21.5% weaker than its strength today, which created an environment in which dinosaurs and humans could thrive. This leads us to the effects of a weaker gravitational system.

<u>**Group discussion:**</u>

1. Did *Origins* establish enough evidence for you to accept the notion that gravity was weaker in the distant past?

2. Depending on how much of the oceans' water came from the Flood, which option seems most plausible and why?
    a. 25% of the ocean water was stored up in the canopy, thus reducing gravity by 10%.
    b. 50% of the ocean water was stored up in the canopy, thus reducing gravity by 18%.

3. Do you already see a connection with weaker gravity and with larger life forms on Earth?

Chapter 2
The Effects of Weaker Gravity on Life

Now that we have established that gravity's intensity can be altered and has been altered to greater intensity over time, and now that we have shown that it was potentially 13.5%—25% weaker before the Flood of Gen. 7, the question is, "How does that effect life?" Why does it even matter what the intensity of gravity was in the past? Well, the human body is a dynamic system; that is, it is adaptable to various changes in the environment as long as those changes are not too large. Since we are talking about the effects of changes in gravity for this chapter, we'll focus on specific adaptations of the body to changes in weight exerted on the body.

To illustrate how gravity affects the weight of a body, consider how a person is weighed. There are two primary methods: pressure-based and mass-based scales. The weight is measured in pounds (lb.) or kilograms (kg), but in physics, the unit of measurement for weight is newtons (*force = mass x gravity = ma = m* (m/s^2). Whether we increase the *mass* or increase the *gravity*, the *force* increases proportionately. And the *force* does not know whether it was an increase in *mass* or an increase in *gravity* that caused the *force* to increase. And for this reason, examples that will be discussed below of the destructive force from being overweight on life prove the destructive force from an increased gravity on life. And conversely, the beneficial force from being at an ideal weight on life proves the beneficial force from a weaker gravity on life. This is a 1:1 ratio.

Different planets have different mass, and this affects the intensity of their gravity and thereby affects the newton weight of an individual. But whether someone stands on the sun, the moon, or Earth, their mass remains the same. However, newtons (or how much someone weighs) will vary greatly.

For example:
A 100-kg man (220 lb.) weighs <u>16.5 kg</u> (36 lb.) on the **moon**. Mass equals 100 kg.
A 100-kg man (220 lb.) weighs <u>99.5 kg</u> (219 lb.) at the **equator**. Mass equals 100 kg.
A 100-kg man (220 lb.) weighs <u>2,804 kg</u> (6,169 lb.) on the **sun**. Mass equals 100 kg.

Review: The greater the gravitational force, the more an object weighs. The weaker the gravitational force, the weaker an object weighs. In both scenarios, mass remains unchanged. Both mass and gravity equally and proportionately affect the force because *f = ma*.

How does the human body respond to varying amounts of load (downward force) placed on the body? When people are overweight, this puts an extra strain on their knee joints, and they have an increased propensity of having arthritic knees, a result of premature breakdown of the cartilage and meniscal material. Or if someone is lifting a heavy object, this increases the risk that he or she could injure a lumbar disc in the lower back. Or when someone is overweight, this usually increases blood pressure and can cause kidney failure, varicose veins, complications of the vascular system—such as heart failure and death, and so forth. This illustrates how extra axial loads placed on the body can cause life to suffer.

An example of the effects of strong gravity on life relates to height. If someone is measured in the morning and measured the same day at night, they will be fractionally shorter at the end of the day. Why? Gravity squishes some of the fluid out of the vertebral disc. This even affects people in the long term. Older people start to shrink fractionally in size. You'll hear older people say, "I used to be 5 feet 11, but now I'm 5 feet 10." Gravity has permanently affected them over the long term. This illustrates that extra axial loads placed on the body can cause living beings to suffer. With a stronger gravitational force, it is harder for a living being to grow to a great height and to live as long as in a weaker gravitational environment.

Since mass and gravity both equally affect the force, then too much gravity restricts the quality of life and the longevity of life; the greater the gravitational force, the greater the strain on the body to survive. Increased gravity makes it harder for the heart to pump blood throughout the body and harder for the body to pump interstitial fluid from the lower extremities up to the torso as well, resulting in clear fluid that pools at the ankles as swelling.

This has a long-term adverse effect on subsequent generations because of mutations in the genome that are passed down to the offspring. For example, an obese people subsequently suffer from kidney failure and mutate their gene as a result. Then their offspring are utilizing a genome with slight mutations in the DNA information that may impair their kidney function. This is not evolution, which adds new information to the DNA code for a new function, or a new kind of life. Some definitions will help us discern the difference between adaptation and evolution. The ability to adapt to changes in the environment is based on the preexisting information in the DNA code. And for this reason, it is not that adaptation produces new information in the DNA that leads to new kinds of creatures, it is that preexisting DNA from the beginning allows adaption and natural selection to occur. So this is not evolution. If the environment causes beneficial new information in the DNA that eventually leads to new functions and new kinds of life—then that is evolution. Outside stimuli can lead to two results: mutations of the DNA code that result in a loss of information, which leads to a loss of function or impaired function, and an emphasis on existing information, resulting in a modification of an external feature, such as a change of species, though the kind remains the same. Both evolution and adaptation are based on outside stimuli. However, evolution relies on mutations in the DNA code to produce new DNA code that produces new functions, which it doesn't, and adaptation is based on preexisting DNA information to achieve modification of features to better suit the environment, which it does. Outside stimuli, such as a series of cold winters, can produce woollier sheep. But the sheep adapt because the information required for adaption already exists in their DNA code, whereas evolution is a change in the DNA code to produce new information that produces a new function or new kind of life. Evolution and adaptation are vastly different; adaptation already has the information to adapt, whereas evolution requires new information to evolve.

Review: Increases in gravity adversely affect the life of humans, including the longevity of life, the height of living beings, and the quality of life.

Whether it is gravity that causes the increased downward force or a person's mass, it doesn't matter because both affect force equally. Force is mass x gravity. Thus, the body doesn't know if the increased force is due to their obesity or increased gravity; it is still an adverse affect. What about a weaker gravitational system? How does that alter life? We can't travel back in time to observe what effect a 13.5% to 25% reduced gravitational system had on dinosaurs, but we can observe the effects that weaker gravitational force has on life today.

Take for example changes in someone's weight: an increased load on the skeletal system mimics stronger gravity and how that effects bone structure. If someone becomes grossly overweight, the body responds by increasing bone quantity (larger in girth, not length) and quality (denser and less porous) to compensate and adapt to the increased load. Heavyset people rarely suffer from osteoporosis (thinning of the bone) as skinny people do as they age. And conversely, a decreased load on the skeletal system mimics weaker gravity. For example, when people grow old, if they are too skinny, they increase their odds of suffering from osteoporosis (thinning of the bone) and their bones being more porous. Why? With changes in gravity or axial loads acting upon the body, the body responds with the skeletal system proportionately modifying its density and quality of bone structure to compensate for the changes in gravity or axial load.

Let's turn to our astronauts for an example. Astronauts residing in space for a prolonged period of time lose a certain amount of bone mass. To determine the effects of zero gravity on the skeletal system, 13 astronauts stayed aboard the space station's orbiting laboratory for six months duration. When they came back down to the earth, they discovered that the average amount of bone mass lost was 14%. Three of the astronauts lost up to 30% of their bone mass. The brain perceives that the body doesn't require high amounts of bone density in a zero-gravity environment, and subsequently the brain signals to reduce the amount of calcium, phosphorus, and magnesium in the bone. The skeletal system adapts (based on existing DNA information) to the reduced gravity and adjusts bone quality, mass, and density to match the reduced load (gravity), and the bone become more porous. After recovering back on Earth, the astronauts gained back all but 2% of the lost bone density, quality, or mass.

Remember, the brain adapting to the change in downward force does not know whether it is a change in mass or gravity; it only perceives a change in force and adjusts accordingly. The weight simulates the effects of changes in gravity. Too much weight simulates a strong gravitational system, and too little weight simulates a weak gravitational system. The brain doesn't know if the extra axial loads are from a stronger gravity or from being overweight, and it doesn't know if less axial loads are from a weaker gravity or from being too thin; it just adapts based on existing DNA information and adds more bone mass or reduces bone mass to compensate. It's the same with the dinosaurs: they didn't know they lived in a weaker net gravity; their brains, utilizing existing DNA information, adjusted the bone quality to match the gravitational strength. And that is why dinosaurs had porous bones.

Review: As a result of changes in gravitational force, the brain alters bone quality, density, and mass to proportionately compensate. Weaker gravity decreases the necessity for strong bones and leads to more porous bones that are less dense. Stronger gravity increases the necessity for strong bones that are denser.

Evolutionary paleontologists believe that dinosaurs evolved into birds. Why? They contend that there are many facts that lead them to this undeniable conclusion. One primary reason is that both birds and dinosaurs have porous bone structures, and both laid eggs. It is true that dinosaur bones are porous, and scientists have determined that if dinosaurs were alive today, their bone structure couldn't support their tonnage in this strong gravitational system. It is simply too porous to handle the axial loads of their tonnage in today's gravity. Since there are no fossil records of dinosaurs evolving into birds, and we don't see reptiles evolving into birds today, then the fact that dinosaurs had porous bone structures indicates that they lived while there was a weaker gravitational force on earth rather than because they evolved into birds. Also, dinosaur eggs were soft and leathery, just as their reptile relatives today. Bird eggs have hard calcium shells. They look the same to the untrained eye, but they have different genetics and texture. It is a mistake to think that dinosaurs evolved into birds because they both had porous bone structures and because they both lay a type of egg, especially when one considers how changes in gravity affect bone quality, and there are zero transitional fossils.

During a diagnostic class in graduate school, my professor said, "When you hear hooves, don't think of birds; think of horses." His point was that one shouldn't overlook the obvious. Just because something is obvious and simple doesn't make it wrong.

Gravity affects our bone density, and dinosaurs had porous bones. Our gravity was weaker in distant past millennia, and there is a natural, obvious link. Dinosaurs had porous skeletal systems because gravity was weaker. This also explains why we don't see large dinosaurs today because gravity has increased over the millennia, not because dinosaurs evolved into birds.

Note that I wrote that there are "no fossil records of dinosaurs evolving into birds." This is important to know. Not too long ago, evolutionists were very excited to discover a fossil that linked dinosaurs evolving into birds. However, it was later proven that the archeologist falsified his data. This infamous scientific hoax of the missing link is called "archaeoraptor." The evidence was falsified, a deliberate attempt to fool the masses. And any dinosaur fossil that has alleged protofeathers representing the burgeoning stages of the evolution of feathers on dinosaurs is merely fraying of scales from the pressure that causes fossil formation in the first place. There will never be a missing link found showing dinosaurs evolved into birds. Why? Because they are different kinds.

Review: Dinosaurs had porous bone structures in accordance with a weaker gravitational force, not because they evolved into birds. *Photo credit: Wikipedia.org/archaeoraptor.*

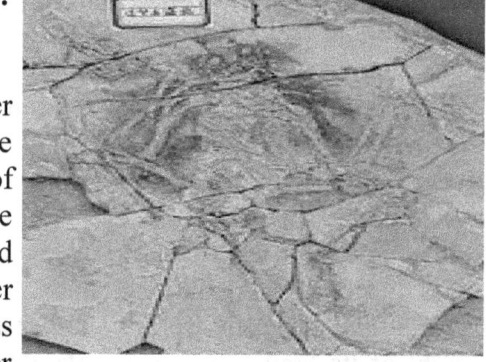

We've discussed that an increase in downward force from either gravity or mass puts extra strain on the body and decreases the quality and quantity of life. Since the negative aspect of increased gravity shortens life, then the positive side of the argument should be valid as well. Well-maintained blood pressure increases the odds of living a long life. A weaker gravity equals reduced blood pressure, and thus there is less strain on the heart and organs of the body. Therefore, weaker gravity causes less strain on the vascular system and prolongs life. And for this reason, a reduced downward force from either gravity or mass puts less strain on the body and increases the quality and quantity of life. Gen. 9:4–6 gives clarity of the significance of blood to life:

> Only you shall not eat <u>flesh with its life, that is, it's blood</u>. Surely I will require your lifeblood; from every beast I will require it. And from every man, from every man's brother I will require the life of man. "Whoever sheds man's blood, by man his blood shall be shed, For <u>in the image of God He made man.</u>"

Life is in the blood. An increase in gravity makes it harder for the heart to pump blood throughout the body. Thus, increased gravity equals a decreased quality of life and quantity of years of life. Weaker gravity would alleviate this problem and make life easier to sustain.

Take fighter jet pilots, for example, to see the effects of gravity. When a fighter jet pilot pulls a large amount of "G" (Gravity) forces, their blood starts to pool around their lower extremities (legs), and they pass out (black out) from a lack of oxygenated blood going to the brain. To combat this, they wear an anti-G suit (pressure suit) to squish the lower extremities and abdominal region. This prevents blood from pooling in the legs and abdomen and helps keep the pilot conscious, preventing the pilot from blacking out. Therefore, the effects of greater gravity make life harder to sustain.

Review: Life is in the blood. Increased gravity makes it harder for the heart to circulate blood, which decreases life span and quality of life. Conversely, weaker gravity makes it easier for the heart to circulate blood, which increases life expectancy.

Does a change in gravity affect the height of living creatures on the earth? Logically, gravity affects height because growth in height is opposite to gravitational force. Therefore, weaker gravity makes it easier to grow taller, and stronger gravity makes it harder to grow taller. But let's see if we can wrap our minds around the concept with some examples.

The largest creature on earth is the blue whale. They have been known to reach 190 tons, and 33 meters (110 ft) in length. No small coincidence that the live in a zero gravity environment—the ocean.

The fastest growing thing on the planet in terms of height is bamboo. It can grow up to 3 feet in one day. One key factor is that bamboo has a thin-walled cortex and a hollow core. This reduces the amount of weight it has to push up against gravity, thereby reducing the amount of force needed to grow. Remember that $m \times g = f$. Therefore, bamboo's reduced mass equals less force and allows it to grow vertically much easier than other heavier botanical life.. When determining force, one multiplies mass x gravity, and the "force" doesn't know if gravity is reduced or if mass is reduced. Either way, if mass or gravity is reduced, then force is subsequently reduced. This exemplifies how gravity restricts height. In other words, if bamboo were thicker and not hollow so that its mass was much heavier, and we grew it in conditions with gravity being greatly reduced so that the amount of downward force were still the same as compared to the real-world scenario, then the height of growth would remain the same. Hence, gravity affects vertical growth.

The second fastest growing thing on the planet in terms of height is kelp. It can grow up to 1.5 feet per day. One of the factors that gives kelp an edge is that it grows in ocean water, and the buoyancy force causes kelp to float up. The bottom line is that weaker gravity equals greater height and accelerated vertical growth. But gravity is not the only cause of accelerated growth—gravity is just a contributory factor. Buoyancy force and oxygen concentration, or the lack of it after the Gen. 7 global flood, is a factor of quality and quantity of life as well.

Take the earth itself as another example. It seems that if gravity can affect earth's terrain and the height of its landscape, then it would definitely affect the living creatures on the earth, as the earth seems less fragile than the life living on the earth. Changes in gravity has altered the shape of Earth, from spherical to ellipsoid (oblong), which means that it is shorter around the poles than around the equator, whereas a sphere has equal distances from every direction.

Since gravity can alter the earth, it can alter the height of something growing on the earth. Since we know gravity was weaker in past millennia, this explains a portion of how dinosaurs grew so large. This is because there was greater buoyancy force at creation, gravity was weaker at creation, and oxygen concentration was higher at creation. The net effect before the Flood was weaker downward force on the dinosaur's skeletal system from gravity. This allowed them to grow to great heights. And this weaker downward force on the dinosaur's circulatory system and organs allowed them to live longer and healthier lives.

Review: Weaker buoyancy force (post-Flood) led to increased gravity, which directly reduced the height and size of life on earth.

Since humans lived 900+ years, and gravity was less intense, then why didn't they grow super tall? Humans and dinosaurs are not the same kinds of creature. Dinosaurs are reptiles, and reptiles continue to grow as long as they are alive. However, humans stop growing shortly after puberty when hormone production causes the epiphyseal plates (growth plates) to fuse, resulting in a termination of vertical growth. Therefore, human growing patterns are different than dinosaur (reptile) growing patterns, and this is/was determined by the DNA.

Dinosaurs living in a weaker gravity environment with a greater buoyancy force, and increased oxygen in the past, before the Flood, allowed them to live hundreds of years longer than reptiles today. And since dinosaurs continue to grow as long as they were alive, this allowed them to reach extreme heights. But humans grow vertically for only a portion of their life. Vertical growth stops when the growth hormones cease production, and this is hardwired in our DNA to be around 17–21 years of age for boys and 15–17 years of age for girls. However, when the endocrine system still produces growth hormones because of some stimulation to the pituitary gland after the epiphyseal plates have fused, then some physical features will continue to grow and protrude, such as with acromegaly, which can result in enlarged hands and feet, a deeper voice, and a pronounced jaw, brow ridges, and cheek bones. In addition, even today, facial bones never stop developing. Thus, imagine what humans looked like if they lived 900+ years—with more pronounced eye brow ridges and jaws. Does this sound like some of those creatures described as Cro-Magnon men, Homo erectus, or hominids that evolutionists talk about? Possibly. There have been so many known fabrications and incidents of falsified evidence of discovering missing link creatures, such as "Lucy," that it's hard to keep up. Therefore, this explains why humans never grew to super heights of say 25–30 feet tall before the Flood. This also sheds light on how humans living in a pre-Flood environment (with weaker gravity and an increased oxygen concentration) would thrive in life and grow to ~10 feet tall.

Review: Unlike dinosaurs, humans stop growing vertically shortly after puberty. However, dinosaurs are reptiles, and they continue to grow as long as they are alive. Imagine if today's reptiles lived 600+ years in a weaker gravity and continued to grow. How large do you think they would become?

The Bible clearly and directly gives the ages of life of the genealogies from Adam to Noah and to Abraham to Jacob/Israel. And a pattern is revealed when studying the ages. The average age of life from creation to the Flood of Gen. 7 is around 900+ years. Then the Flood comes, and the ages dramatically start to drop off. Look at the list of people who lived after the Flood. They are in chronological and genealogical order:

Noah lived	950 years (Gen. 9:29).
Noah's son Shem lived	600 years (Gen. 11:10–11).
Shem's son Arpachshad lived	438 years (Gen. 11:12–13).
Arpachshad's son Shelah lived	433 years (Gen. 11:14–15).
Shelah's son Eber lived	464 years (Gen. 11:16–17).
Eber's son Peleg lived	239 years (Gen. 11:18–-19).
Peleg's son Reu lived	239 years (Gen. 11:20–21).
Reu's son Serug lived	230 years (Gen. 11:22–23).
Serug's son Nahor lived	148 years (Gen. 11:24–25).
Nahor's son Terah lived	205 years (Gen. 11:32).
Terah's son Abram (Abraham) lived	175 years (Gen. 25:7).
Abraham's son Isaac lived	180 years (Gen. 35:28–29).
Isaac's son Jacob lived	147 years (Gen. 47:28).
Jacob's son Joseph lived	110 years (Gen. 50:26).

The Bible does not directly say that the ages of mankind's life span dropped because of the Flood or because of the resulting decrease in buoyancy force, oxygen levels, or the increase in gravity. We have to use deduction to put the pieces together. And here are the pieces: Before the Flood, there was a canopy of salt water surrounding the atmosphere, which resulted in global ambient temperatures and no deserts, no polar ice caps, smaller seas, and no oceans. This led to massive vegetation, which led to high oxygen production, increased atmosphere density, and an increase in heavier oxygen elements. The canopy of water resulted in an increased weight of the atmosphere, which led to an increased buoyancy force, which led to a reduced weight of all objects by affecting gravity's net force. With net gravity weaker, this allowed living beings to live longer, and they grew taller and larger.

Then a catastrophic event, the global flood of Gen. 7, came roughly 4,500 years ago (about 1,650 years after creation—based on genealogies), resulting in the loss of the canopy, which was one source of the 40 days and nights of rain. During the Flood, there were very high temperatures from asteroid impacts and from water bursting out of the crust of the earth as the fountains of the great deep. This was the breakup of the crust of the earth and resulted in hundreds of massive volcanoes and fast tectonic plate movements. This catastrophic event was a global flood, and the floodwater was mixed with sand, silt, rock, clay, dirt, soil, vegetation, biomass, and so forth, and this caused the formation of oceans and the layers of soil on the crust and the formation of all fossils, petroleum, petrified rocks, coal, ice ages, and polar ice caps. Oceans limit vegetation growth, resulting in reduced O2 production. Losing the protection of the canopy of water meant the beginning of deserts, which also limit vegetation growth and reduce O2 production levels. The formation of polar ice and equatorial heat caused the initiation of wind, the hydrological cycle, rainbows, clouds, and rain. The loss of the canopy of water allowed solar energy, x-rays, cosmic rays, and gamma rays to penetrate to Earth's surface and increase harmful genetic mutations that result in diseases and reduce the quantity and quality of life. Also, this resulted in decreasing the buoyant force and increasing net gravity. Mankind had to adapt to the changes and subsequently lived shorter lives. The adaptation of life to these changes caused a reduction in life span and the size of all living creatures, some of which continued to grow as long as they were alive, such as the dinosaurs.

Back to our discussion, with the reduction of ages from 900+ pre-Flood to <100 post-Flood as each subsequent generation adapted to life after the Flood, the effect of a reduced buoyancy force and the subsequent increase in net gravitational force and reduction of oxygen concentration, not only affected longevity of life for humans, but also every living thing. Within 13 generations after the Flood, the life expectancy was reduced from 900+ years of life to 110 years of life. By the way, this wasn't new DNA information, this was adaptation because of existing DNA. Mankind will once again return to the same longevity, and it will be based on the DNA code that allows life to adapt to a favorable new environment and live 900+ years again. We have slightly different DNA than Adam and Eve because we are copies of copies, and we are a mixture of our parents for hundreds of generations. But within each human exists DNA information to once again live long lives as Isaiah 65:20 prophecies.

Something changed to cause this significant reduction of life span. And that something that changed was the loss of the canopy of salt water coming down to Earth, which reduced the buoyancy force, causing a net gravity increase and blocking vegetation growth, which decreased the oxygen concentration in the atmosphere. This resulted in a strain on living beings to survive. This strain from increased gravity and decreased oxygen caused the life span of all living creatures to be reduced. In addition, with reduced life spans, this limited creatures that typically continue to grow as long as they live to decades of growth instead of nearly a millennium of growth. Therefore, creatures that share similarities with dinosaurs and prehistoric creatures, such as the crocodile, alligator, Komodo dragon, great white shark, and tiger shark, are still living amongst us, but they are just much smaller than when their ancestors once lived and thrived in an environment with weaker net gravity, increased buoyancy force, and increased oxygen concentration.

Review: The Bible is clear that humans lived 900+ years before the Flood, and their life spans gradually declined. This is clearly a result of the Flood of Gen. 7 and the changes caused by reduced buoyancy force, increased gravity, and decreased oxygen concentrations. An obvious link is that since human life span declined, then so too did all creature's life span.

The Bible explains that life on earth was adversely affected by sin. And this onset of sin on earth started a chain reaction of judgments on everything created so that every creature groaned and suffered and longed to be free from the bondage of sin. Romans 8:18–25:

> Creation eagerly awaits for the revealing of the sons of God. For the creation was subjected to futility, not willing, but because of Him who subjected it, in hope that the creation itself also will be delivered from the bondage of corruption . . . the whole creation groans and suffers the pains of childbirth together until now.

The whole of creation will be free from the bondage of sin again one day in the future, just like it was at creation. Does this mean that oxygen concentration will be higher again and net gravity will be weaker again and buoyancy force will be stronger? Yes. Will the canopy of salt water surround the atmosphere again? Yes. How? A cataclysmic event is scheduled to occur soon, which we know from Revelation as seals, trumpets, and bowls, a series of judgments of catastrophic proportions. This will include huge asteroids crashing into the earth. So severe will the impact be that the earth will split asunder and wobble like a drunkard, like a slow spinning top. These huge asteroid impacts cause the earth to split, and the ocean waters pour into the super hot magma; this expels all of the ocean water back up into space—where they were before the Flood of Gen. 7, and this results in the earth spinning faster from the water shooting up, acting like a jet engine, increasing the spin of Earth back to the original creation speed, which means the day will be approximately 17 hours long again. This restores the canopy and the environment back to creation days.

Is this physically possible? Very much so. Where does the Bible talk about this event? Isaiah 24:18–19 says, "The windows above are opened, and the foundations of the earth shake. The earth is broken asunder, the earth is split through, the earth is shaken violently. The earth reels to and fro like a drunkard and it totters like a shack." And Revelation 16:17–21:

> There was a great earthquake, such as there had not been since man came to be upon the earth, so great an earthquake, so mighty. The great city was split into three parts . . . every island fled away, and the mountains were not found. And huge hailstones, about 100 pounds each, came down from heaven upon men.

A severe impact from a large asteroid could split the earth, and then the oceans would pour into the superheated mantle inside the earth. When the earth splits asunder, the hot iron core is exposed, and all the oceans drain into the magma and explodes up into space. As the ocean water is being propelled into space, our earth's gravity would keep the water in orbit surrounding our atmosphere, and this would be the beginning of the reformation of the original canopy of salt water that existed before the Flood. This also explains how 100-pound hail stones could exist in the Revelation prophecy; as the oceans' water was being propelled up into space, some of that water would just reach high in the atmosphere and come back down as 100-pound hailstones.

This would result in a new Earth, where there would be no ocean and a new heaven. Imagine how differently heaven would look with the canopy of water reestablished, creating a spherical canopy that would eventually morph to an arch in the sky from lunar gravity and centripetal force. Revelation 21:1: "Then I saw a new heaven and a new earth; for the first heaven and the first earth passed away, and there is no longer any sea."

What, no seas, no oceans? Correct. Isaiah 24:18b–20 and Revelation 16:17–21 are sequential and parallel verses that go together. But most importantly, all that water shot up into space will reconstitute the canopy of water again. This will increase buoyancy, which will reduce gravity, allow almost 100% of the land to be useable for vegetation, increase oxygen back to creation conditions, and cause life to thrive again. Human beings will return to living 900+ years. Essentially, everything returns to a pre-fallen environment—creation.

This doesn't mean there will be no water on earth. Remember that there will be ample fresh water flowing out of the two rivers coming out of the Temple Mount, which causes the northern portion of the Dead Sea to heal from the east and flow into the Mediterranean from the west (Ezekiel 47). With that much water flowing out of the Temple Mount, there will be freshwater lakes, not salty oceans.

Chapter review: Human beings once lived 900+ years, and that day will return again when Jesus comes and sets up His earthly kingdom to reign for a thousand years. Increased gravity adversely

affects living creatures. The increased net downward pull of gravity has reduced the rate of growth, the height of living things, and life expectancy. Increased gravity even affects the terrain of the earth and the shape of the earth, changing it from spherical (like a ball) to ellipsoidal (a ball that is slightly fatter at the equator and squished at the poles). There is no doubt that the Flood of Gen. 7 caused a reduction of buoyancy force, increased gravity, and decreased oxygen, which adversely affected life on earth. Dinosaurs didn't evolve into birds; they lived in weaker gravity and stronger buoyancy, which allowed the dinosaurs to have porous bone structures because skeletal structures become thicker with increased gravity and become more porous with a decrease in gravity.

Group Discussion:

1. What seems more likely, that dinosaurs evolved into birds and proof of this is their porous bones, or that dinosaurs had porous bones because gravity was weaker when they roamed the earth?

2. Imagine that you were able to view a time-lapse film of Tyrannosaurus Rex living to 500 years of age in a weaker gravity environment with high oxygen concentration. Then oxygen levels gradually declined and the strength of gravity gradually increased, and you were able to witness many generations of T-rex pass. How do you think T-rex's offspring would adapt?

3. The Bible prophecies that creation returns to the conditions of the Genesis creation, but the process to get there is quite destructive. How does reading that the earth splits asunder and wobbles like a slow spinning top challenge you?

Chapter 3
The Canopy of Salt Water

In the beginning of time—when things began—our atmosphere, gravity, and Earth were slightly different than today. How were they different? There were higher oxygen concentrations in the atmosphere, a greater buoyancy force, weaker net gravity, more usable land for vegetation, no oceans, no polar ice caps, no deserts, no four seasons, and no clouds to produce rain. We will explore each of these items and explain how they are not only possible, but probable, and we will provide the evidence to uncover hidden truths.

There is one thing that can cause all the above listed conditions to exist; it is the canopy of salt water that surrounded the earth's atmosphere. The canopy existed from the second day of creation to the global flood at ±2,350 BC. The cross sections below illustrate that the canopy was initially spherical around the atmosphere and morphed over time to a disk-like shape. The proportions and time line are not meant to be exact, only to provide a visual of the hypothesis.

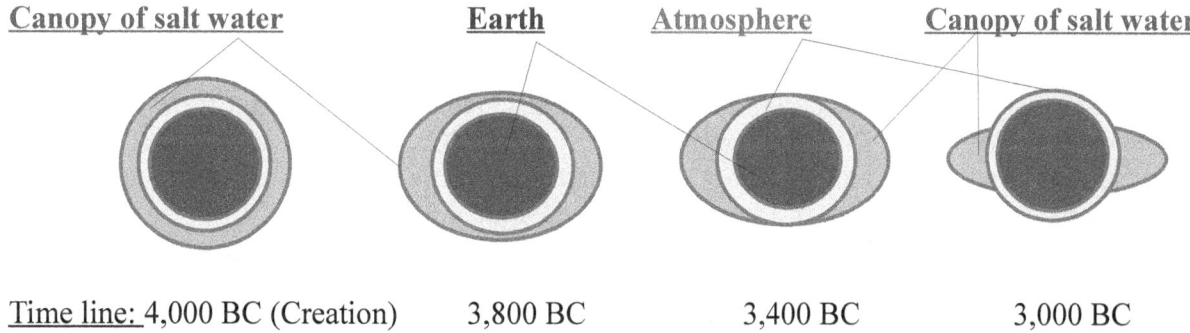

Time line: 4,000 BC (Creation)   3,800 BC   3,400 BC   3,000 BC

This chapter will focus on the canopy of salt water that surrounded the atmosphere from creation until the global flood, which changed the environmental conditions on earth and resulted in increased gravity, decreased buoyancy force, and decreased oxygen concentration and subsequently resulted in a reduction in the longevity of life and the size of life forms.

In the beginning of life on earth, there was a canopy that surrounded our atmosphere that initially hovered spherically around the planet and morphed over time from the lunar gravitational pull and centripetal force to thicken at the equator and thin at the poles, until a visible arch could be seen in the sky. How do we know that there was a canopy surrounding our atmosphere? First, we must start from the beginning before the earth and galaxies were finished forming. In the beginning, God created (from His essence) all the matter in a confined space. At this moment, water and dirt were tightly packed. In the darkness, all the future stars, planets, and moons, were coalescing (gathering matter), and this caused heat. On the outside of the small universe, away from the hot coalescing future stars, there was an external shell of thick ice; this sealed the universe for internal pressure, heat, and charge to generate. The combination of heat and pressure allowed for water to exist in liquid form without a star to warm it, and the earth was shapeless and void of life (Gen. 1:2). These were the first hours of the first day of creation. Let's focus on the water for a little bit.

Where was the water located? The water was located throughout a small universe and in our even smaller solar system. At this point in Gen. 1:2, there was nothing with form, no light, and no life, and the universe was squished to the size of many galaxies. Gen. 1:2 says, "The earth was formless and void, and darkness was over the surface of the deep, and the Spirit of God was moving over the surface of the waters." Why is *waters* plural? This could refer to the four forms that water exists in: solid, liquid, gas, and plasma. Or this could refer to the massive amount of water. Probably, the answer is both. In the beginning, the earth, sun, moon, and stars, were not finished forming yet. They were all spinning and coalescing matter. This looked like trillions of hurricanes of hot matter gathering matter, with water filling the gaps in between each of them. The word *deep* is used to refer to abundant waters in the Pentateuch (Bible books 1–5). How do we know this? Gen. 7:11 cites the "fountains of the deep." (Strong's Concordance #8415: *t(e)howm*: an abyss (as a surging mass of water), main sea or deep subterranean water supply, deep place, depth.

Let's determine how water could exist in four forms before the formation of the stars. There is a simple twofold solution to this problem. The first is salt and minerals. Water freezes at 0°Celsius / 32°Fahrenheit when it is purely H2O. But if we add salt, this will lower the freezing point of water to -21°C / -5.8°F. This is why cities distribute salt on the roads in the winter.

Before the fall of mankind, there was no death, as death came as a result of sin (Romans 6:23). The Dead Sea could not have existed before the fall of mankind because nothing lives or grows in the Dead Sea because it is too salty. It is an example of death. Therefore, the Dead Sea was formed after the fall of mankind and most likely from the global flood of Gen. 7. My contention is that the Dead Sea and some of the saltiness of the oceans and the vast shallow salt deposits underground were formed as a result of God's judgment of sin, with the Flood of Gen. 7.

The deeper salt deposits were probably formed from the formation of the hot Earth while it was surrounded by salt water (day 1 of creation). As a hot Earth coalesced matter, it was surrounded by water (2 Peter 3:5). This environment caused evaporation, which separated fresh water away from the salt; this resulted in large deeper salt basins that were pure salt. This also resulted in fresh water being closest to the earth's surface and the saltiest water being farthest away from the earth during the first two days of creation. And that is probably how God separated fresh water as the fountains of the deep and salt water as the canopy above the atmosphere and formed the deeper salt mines.

It is interesting to note that evolutionists believe that a mixture of complex chemicals rained down upon the rocks and dirt to form a primordial pool of complex chemicals; this pool spawned life. I hypothesize that a mixture of complex chemicals rained down upon the earth as well during the global flood of Gen. 7. This also could explain how large aquatic dinosaurs could have died off from a global flood. Potentially, some great sea creatures were not able to adapt from a freshwater environment or 1% salty seas to the saltwater conditions of the newly formed oceans (3.5% salt concentration) after the Flood. Some adapted and survived, and some couldn't and became extinct. During the Flood, there were so many volcanoes erupting and asteroids hitting the earth that actual trauma or acid waters could have killed most creatures as well. How did the salt get collected together in some locations? Toward the end of the Flood saga, as temperatures dropped from intense heat to freezing from 40 days of no sun and a globe covered in seawater for 150 days, ice formed. When the ice formed, it was fresh water, thus leaving behind salt to saturate the waters where the ice formed. With those areas saturated with salt, no more salt could be diluted in solution; thus, it descended to the bottom in solid form and piled up where the ice was forming. With turbulence and swirling waters from the caverns of the deep that burst open and from when Pangaea broke apart, this explains how large deposits of salt mines coalesced and how dinosaur graveyards of bones pooled together.

Marine life is sensitive to the salinity of water. Change the salt content too much, and marine creatures will die. Potentially, the seas that were formed by God in the beginning on the third day of creation were only 1% to 2% salt concentration. And the ice age, by freezing fresh water out of the salty floodwater, increased the oceans salt concentration to 3.5%, which killed off some water-dwelling creatures that couldn't handle the sudden increase in salt.

Review: Before the earth and atmosphere were formed, water and dirt existed; the dirt was formless, and no life existed. The earth was still coalescing matter together as it spun. The water existed in four forms (liquid, solid, gas, and plasma) and was very abundant. Some water remained in a liquid form partially because of the salt concentration. This is where all the salt

came from on the earth.

We just explained how salt helps water stay in a liquid form despite very low temperatures, but that theory only takes care of temperatures dropping to -21°C. What about before the sun and stars were finished forming, that is, on the first three days of creation? Space gets much colder than -21°C. The answer is pressure, friction, electrical charge, and heat. At this point in time (before the sun, moon, and stars had completed their gathering of matter), the universe was formless, lifelessness, and densely compact, and trillions of future stars were spinning, coalescing matter, and glowing hot by the middle of the first day until they ignited on the fourth day. The universe was much smaller in size than the vast spread-out size of our current universe. How do we know this?

For one thing, science has determined that the universe is still expanding and appears to be expanding at an accelerated rate. The other way of knowing the universe is expanding is that the Bible explains that God spread out the universe (Genesis second day, and Isaiah 40:22), which means that it was densely packed at one time. High pressure allowed for liquid water before the completion of the formation of the sun, moon, and stars. High pressure reduces the freezing point and generates heat.

Changes in <u>pressure alter the freezing point</u>, the boiling point, the vapor point, and the plasma point of H2O. With increased pressure exerted, the boiling point of water is higher and the freezing point is lower. For example, an ice skater's blades are said to instantly liquify the ice when the blades touch the ice due to the high amount of pressure. The universe just needed to be dense enough to generate some heat because <u>pressure would have also lowered the freezing point of water well below -21°C</u>. However, since *waters* is plural in Gen. 1:2, then there was enough pressure and heat closer to coalescing stars to not only keep water in liquid form, but also to turn water into vapor and eventually for plasma to generate the fusion of hydrogen by the fourth day of creation for stars. This pressure and heat wanted to expand, but couldn't because the outside shell of this small universe was solid frozen ice that wasn't near a heat source. This closed, sealed system called the universe, built up heat, pressure, and electrical charge as trillions of spinning balls of matter (eventual stars) coalesced into spinning magma (liquid rock and metal) that each started to glow hot from God saying on the first day, "Let there be light." Heat, charge, and pressure grew to unsustainable proportions for the solid ice outer shell, and this caused trillions of simultaneous Big Bangs (on the second day); this caused cavitation (sudden release of pressure that forms air bubbles) of the water surrounding every planet, forming atmospheres.

To have lots of pressure in a small universe, it had to be densely packed. Maybe our entire universe was compressed into the size of a couple of hundred galaxies worth or even a thousand galaxies worth. There is no way to know for sure because it depends on how much matter (stars, moons, and planets) exist in the universe. It's not logical for the entire universe to have been compressed into the size of a dot—or a singularity—as Big Bang theorists hypothesize. It would require too much energy, intelligence, and design, to organize and compress all matter into an area the size of a dot as energy, and that much generated heat would vaporize all matter. Big Bang theorists have thus shifted their concept such that—at the moment prior to the Big Bang—there was only energy that was compressed down to the size of a dot and zero matter. All the matter in the universe was compressed together so tightly that all matter existed only as energy and not as matter. In physics, matter and energy really are the same thing, so this hypothesis is acceptable. However, the singularity hypothesis is only theoretical and not observed and has not been tested. What has been tested and observed via the Hadron Collider is that trace amounts of matter can be created from energy and intelligence. Since energy and matter are the same thing in the physics world, then really, what they have achieved is converting huge amounts of energy into trace amounts of matter. The Higgs boson particle establishes that massless particles can collect trace amounts of mass as they pass through a field of protons that have just collided into each other near the speed of light. When I say trace amounts of mass, I'm talking about the mass equivalent of 125 gigaelectron volts, which is 2.29×10^{-22} g. To put that in perspective, the average raindrop weighs about 2.0×10^{-1} g, and a grain of sand weighs 4.4×10^{-3} g. Physicists have used energy to form mass that is 1.0×10^{-19} times smaller than a grain of sand and with an alarming amount of energy and intelligent design required to do so. To give an indication of the amount of intelligence involved, it is estimated that >10,000 scientist and engineers were involved in the project. Thus, they have a long way to go to form something as large as a grain of sand and an infinitely longer way to go to account for all the matter in the universe. What they have established is that it takes intelligence to achieve what they have done—not random chance. Their work, intended to establish natural means of forming matter and to suggest that there is no God required, has only firmly supported Intelligent Design. And the large amount of energy/power required to collide two massless particles into each other and form miniscule amounts of matter establishes that only an all powerful God could have created the amount of matter that exists in the universe.

Back to our discussion, there are some ice volcanoes on Saturn's moon Titan. Cosmologists have determined that as a result of Titan's orbit around Saturn, the gravitational forces have built up friction and melted water below the surface of Titan. Think about that for a minute. Titan is too far away from the sun to be warmed by the sun's heat, and it doesn't have an atmosphere to contain warmth. Yet, because of tidal friction, it has liquid below its surface. This is similar to the principle of the canopy having a liquid interior, perhaps caused by friction as on Titan or from the moon's gravity causing tides within the canopy itself and from pressure generating heat.

Review: Before God stretched out the universe, it was densely packed, and this resulted in high pressure, friction, heat, and electrical charge. This kept water in a liquid state at the center of the compression and frozen on the outer shell. This correlates to the middle of the first day of creation.

When was the canopy of salt water formed, and how was it formed?
On the second day of creation (Gen. 1:6–10):

> Then God said, "Let there be an expanse in the midst of the waters, and let it separate the waters from the waters," God made the expanse, and separated the waters which were below the expanse from the waters which were above the expanse; and it was so. God called the expanse heaven. And there was evening and there was morning a second day. Then God said, "Let the waters below the heavens be gathered into one place and let the dry land appear"; and it was so. God called the dry land earth, and the gathering of the waters He called seas.

Later on, God places birds in the lower heaven—our atmosphere. The text is clear that there is water below and above the atmosphere. If this isn't clear enough to show that water existed above the atmosphere, consider Psalm 148:4, which says, "Praise Him Highest heavens, and the waters, that are above the heavens!" And 2 Peter 3:5 says, "The earth was formed out of water and by water."

The creation of the atmosphere was a result of God utilizing natural and supernatural processes. For example, He supernaturally created matter at the beginning of the first day, and He did so with rotational motion already established for each celestial body, and they were coalescing matter at their birth. With enough mass coalesced, the spinning matter began to glow hot from magma at the middle of the first day. Pressure, heat, and electrical charge continued building until the frozen exterior shell of the small universe could not contain the immense internal forces, and trillions of Big Bangs occurred at the beginning of the second day of creation. This rapid expansion of the universe caused cavitation and the instantaneous release of pressure from the liquid that surrounded Earth, and this caused our liquid atmosphere to instantly turn to gas at the middle of the second day. The process of cavitation is seen every time a new bottle of soda is opened and the bubbles form and come out of the pressurized liquid.

After God created the atmosphere and the canopy of salt water hovering around the atmosphere, what kept the water suspended in space and hovering around the atmosphere? Why didn't the canopy come crashing down upon the earth immediately? There were five contributing factors: distance, buoyancy force, the ice/arch support, the spin of the earth, and covalent bonds.

Distance: The greater the distance from earth, the weaker gravity becomes. The following is the acceleration due to gravity at various altitudes:

Altitude	Gravity (m/s/s)
0	9.81
62.5 km	9.67
125 km = 77.5 miles	9.50* 3.2% weaker
1,000 km = 621 miles	7.33

One of the ways of keeping the canopy hovering around the atmosphere is simply by placing it at a certain distance from the earth's gravitational pull. *Image credit: World-builders.org. Charter College, Cal. St. Univ., Los Angeles.*

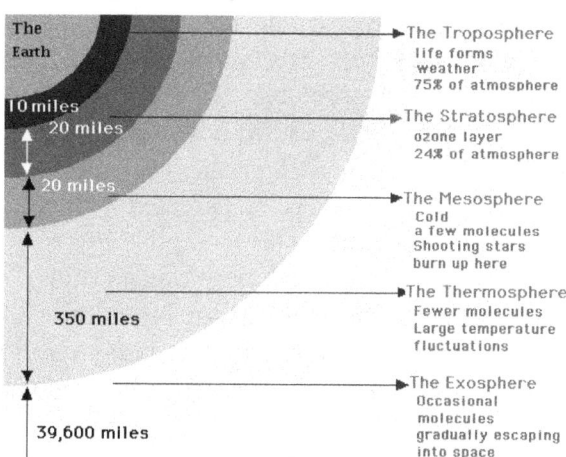

The **exosphere** extends ~640—64,000 km (~400–40,000 mi) above earth. This is too far away for the canopy to cause buoyancy force.

The **thermosphere** extends ~80–640 km (~50–400 mi) above earth; solar radiation absorption here = 230°C (440°F). Solar radiation warmed the inner fluid inside the canopy as well.

The **mesosphere** extends ~50–80 km (31–50 mi) above the earth; at -90°C (-130°F), space debris burns up and shooting stars blaze. A possible location of the canopy to compress the atmosphere down to this altitude, exert buoyancy force, and remain elevated until the global flood.

The **stratosphere** extends ~16–50k m (10–31 mi) above the earth; this is a protective area where the ozone layer floats and 24% of the molecules in the atmosphere exist; here, the air gets warmer. Another possible altitude for the canopy—here it would exert greater buoyancy force, as the atmosphere would be denser.

The **troposphere extends** up to 16 km (up to 10 mi) above the earth; it is where weather forms. It contains 75% of the molecules in the atmosphere. This seems too close.

If the canopy was ~600 miles from Earth, gravity would be reduced at this altitude. However, the canopy wouldn't need to be 600 miles away from Earth, but maybe in the mesosphere around 40–50 miles from earth's surface; at this altitude, it would reduce gravity by only 1.5%. This isn't much by itself, but it doesn't have to be much to just contribute.

Buoyancy Force: The pressure exerted from water pushed down upon the atmosphere, causing the atmosphere to be compressed. This increased the density of the atmosphere and increased pressure, which created a buoyancy force that counteracted gravity for all occupants on earth. A principle in physics can be applied to reveal a mystery of the suspended canopy. It suggests that for every action, there is an equal and opposite reaction. Therefore, as the canopy pushed down on the atmosphere, there came a point of homeostasis when the atmosphere pushed back up upon the canopy with the same force. Thus, an equilibrium occurred in which the weight of the canopy was equal to the pressure of the atmosphere's push upwards; thus, the atmosphere held the canopy in place by the buoyancy force.

Ice: During the first day of creation, the water surrounding the earth would be all liquid from the heat, pressure, and salt, encompassing the earth with no air surrounding the earth. But after the Big Bangs, the sudden release of pressure resulted in a rapid expansion of the universe to form the atmosphere—via the process of cavitation; then the outer and inner portions of the canopy would be subject to freezing temperatures. Thus, the outer portion of the canopy of water would be frozen almost instantly, except the central portion of the canopy would still be liquid. Since the canopy of water was spherical on the second day of creation, with a frozen exterior and inner circumference (and an internal liquid reservoir), it would act as a support bridge holding itself up. An example of this is an igloo. Eskimos use an arch method and build a house completely out of ice with no other supports, and their igloos can withstand gravity and are stable through severe arctic blizzards. When a crack occurred near the equator because of tidal expansion from the moon and sun, the liquid inside would fill the crack and be resealed via the cold. Thus, the spherical canopy morphed toward a bulge at the equator with each rotation of the earth and lunar tide. And the arch support was strong enough to solely support the canopy above the atmosphere without any additional factors. Architects will tell you that an arch is one of the strongest structures.

Spin of the earth: The other means of keeping this canopy hovering around the atmosphere would be the spin of the earth.

Our atmosphere was created on the second day of creation, with water above and below, and it was already rotating. This is implied by the phrase in the Bible for the end of each day of creation: "There was evening and there was morning," and by "x" day. The rotation of the forming earth would keep the atmosphere and canopy rotating with the earth. What caused both the atmosphere and canopy to rotate with the earth? Friction and gravity. The contact of the earth with the atmosphere and the contact of the atmosphere with the canopy would cause a process of near harmonious rotation. Currently, the earth's gravity prevents our atmosphere from escaping into space. It would have also prevented the water from drifting off into space. As the atmosphere spun with the earth, so too would the canopy rotate with the atmosphere and with the earth's rotation. An example of this would be to take a cup of water and start rotating the cup. Eventually, the water in the cup will spin with the cup. This is because of friction with the glass and the water's covalent bonds.

The earth is spinning 1,037 mph on the surface today, but back 6,000 years ago, the length of a day was ~ 17 hours, which means that the earth spun 1,600 mph+. Currently, the entire depth of the atmosphere is estimated at 1,000 km, or 621 miles. It's unknown as to where our atmosphere actually stops because the higher up in altitude, the fewer air molecules there are, and there is no defining edge. We'll have to play around with the numbers and give it our best estimate. With the canopy hovering at an altitude of 1,000 km (621 miles), gravity at this altitude is very weak at only 0.13 m/s/s (roughly 1.3% of surface strength). And considering the spin produces centripetal force that tends to counteract gravity, and presuming a 17-hour day, this results in the canopy of water flying off into space because the centripetal force is greater than gravity. Therefore, the canopy of water could not be that far away from the earth. Let's try an altitude of 62 miles above the earth's surface. With the **canopy hovering at an altitude of 100 km (62.1 miles)**, and factoring the centripetal force from a 17-hour day, this results in **gravity reduced by 4%**. This is not enough by itself to prevent the canopy from raining down upon the earth, but it does contribute.

With the spherical canopy spinning with the earth, this would cause the shape of the canopy to morph toward a disc shape. Initially, on the second day of creation, the canopy would be a perfect sphere, but with each passing day, and with each tidal friction from the moon, cracks would occur in the exterior shell of the canopy closest to the moon; the liquid interior water would fill those gaps, and the spherical canopy would gradually morph to an equatorial bulge. By the sixth day, at the creation of Adam and Eve, there may have been enough thickening of the equatorial bulge to form a visible arch in the sky from water accumulation. This process continued for each planet in the universe, until some formed rings like Saturn; some lost all that water, and one in particular (the earth) maintained an arch in the sky made of water just in time for the Gen. 7 flood.

Covalent Bonds: These are the bonds on a molecular level that bind the water molecule H2O together. The "H" stands for hydrogen, and the "O" stands for oxygen; they are bonded together and held together by mutual attraction and shared electrons in the same valence (the orbit of electrons from the nucleus) called covalent bonds. This isn't the strongest force in the universe by any means, but it does contribute to keeping water connected together. This force would have helped solidify the frozen archway. We learn in chemistry that H2O has a relatively strong bond, and with that strong bond, we can make water flow up against gravity by water's connective properties of bonding. Thus, with the instant formation of ice on the outer side of the canopy, if there was any inner liquid portion of the canopy, there would be an affinity to cling tightly to the solid support arch of the ice and thus suspend the entire canopy in space surrounding our atmosphere. Then a layer of frozen ice closest to the earth sealed the inner liquid water.

No one knows for certain what altitude the canopy hovered at because we weren't there to observe this. What we do know is the contributing factors that played a role in causing it to reach an equilibrium of altitude above the atmosphere, and it doesn't violate any laws of science. When adding the sum of contributing factors from the centripetal force, the altitude, and the buoyancy force from the atmosphere pushing up on the canopy to the covalent bonds holding water together, the frozen external archway of the canopy surrounding the outermost atmosphere holding it up, we see the plausibility of the canopy hovering above the atmosphere.

Review: The canopy of salt water was held in place at its designated altitude by its distance from earth, the earth's spin, a frozen arch support, the buoyancy force of the atmosphere, and the covalent bonds of water itself.

Before God stretched out the two expanses, the first for the universe and the second for our atmosphere, the universe was very small compared to its current size. However, the Big Bang theorists take this smallness to mean the size of a dot. This dot where all the matter in the universe was squished together as exclusively energy is called the singularity. Let's discover if the construct of the singularity of Big Bang theorists is supported or rejected by the Bible.

The reasoning behind our universe not being smaller than the size of several galaxies or our solar system and certainly not being small like the singularity that Big Bang theorists believe in is based on the Bible and science. Let's work with Biblical logic first; God stretched out the universe and then our atmosphere on the second day of creation before finalizing the earth because *heavens* is pluralized in Gen. 1:9. Our universe couldn't be smaller than our solar system because there would be too much heat generated from the intense pressure of a singularity, and the waters mentioned in Gen. 1:2 would not be waters, but vaporized. But the fact that the Bible uses the plural *waters* in Gen. 1:2 before the universe was stretched out and before the sun, moon, and stars were finished forming indicates that the universe was small enough to generate pressure, small enough to generate enough heat to perfectly maintain water in liquid form, but not too compressed that too much heat would be generated for water to be completely vaporized—which the singularity would do. It is safe to say from a Biblical perspective that all the matter in the universe could not be squished smaller than the size of our current solar system, or else water wouldn't be able to exist because there would be too much pressure and heat.

Essentially, if someone believes in the singularity construct of the Big Bang theory and believes in God, they are mutually exclusive from this logic alone, as the Big Bang theory suggests that all matter in the universe was compressed into a tiny dot as energy before the Big Bang. Yet the Bible declares that before God stretched out the universe on the second day of creation (i.e., the Biblical Big Bang), the waters and dirt existed from the first day. Therefore, a theistic Big Bang theorist rejects what the Bible declares as a small enough universe and accepts what evolutionary cosmologists suggest was an extremely tiny (a dot) universe. And why would anyone reject the Bible's record of the beginning for man's hypothesis?. That's right, believers in God are rejecting God's testimony and accepting a version of the beginning from atheists, who at the core, form hypotheses to disprove God. The Biblical small universe was small enough to generate heat to maintain water in liquid form, but not too small as to vaporize all the water and dirt in a singularity. By annulling some words or tampering with the fidelity of the Genesis creation account, or by saying something to the affect that the Genesis creation account is not literal, this opens up a whole can of worms as to what is literal and figurative. Then, when does one arbitrarily jump into the Bible and start reading it literally. Just let God keep what He said in Exodus 20 and Exodus 31 as the definition of the Genesis creation account: that He made everything in six days and rested on the seventh. He says this twice and writes it in stone twice for Moses. Furthermore, God said in Numbers 12:4–8, "If there is a prophet among you, I, the LORD, shall make Myself known to him in a vision, I shall speak with him in a dream. Not so with My servant Moses . . . With him I speak mouth to mouth, even openly, and not in dark sayings, and he beholds the form of the LORD." God is saying that He spoke clearly to Moses face to face just like friends would talk and not in veiled dreams or visions. Therefore, when Moses wrote clearly about the Genesis creation account, his information was based on clear information that God gave him to write, not obscure dreams or visions. Thus, we should interpret it in the same fashion, as clear, straightforward language.

But let's not throw out the Big Bang theory entirely because the Bible teaches that our vast universe, with all the galaxies, stars, sun, and moon, was a densely compact universe at one time. The Bible explains in several verses that God spread out the heavens. Isaiah 40:22 says, "It is He who sits above the circle of the earth, and its inhabitants are like grasshoppers, Who <u>stretches out</u> the heavens like a curtain and <u>spreads</u> them out like a tent to dwell in." And when the Holy Spirit moved as God spoke, the Hebrew word for Spirit is *ruwach*, and it also means there was violence, almost with the intensity of anger, blast, tempest, and spirit (Strong's Concordance #7307). Those definitions indicate the violence of the Big Bang as God stretched out the heavens. There are three heavens: the atmosphere, the universe, and the Throne of God (2 Corinthians 12:1–4). When did God expand or stretch out our atmosphere? The second day of creation (Gen. 1:6–8). When did God stretch out the universe? On the second day as well, but prior to the expansion of the atmosphere. One reason we may discern this is because *heavens* is plural at the beginning of the third day, and the sun, moon, and stars are formed in an already-existing universe on the fourth day of creation (Gen. 1:14–19). Also, physics explains that when there is a sudden release of pressure, such as on the second day of creation, then there is an instantaneous conversion of liquid to gas; that would be our atmosphere. This is illustrated every time we open a soda bottle. The release of the cap (expansion of the universe) violently converts liquid to gas because of a sudden reduction in pressure.

Have you connected the link between the Hebrew word for firmament/expanse in Gen. 1:6 with the word *arch*? One of the definitions that no one seems to be familiar with when God inspired Moses to write the word for firmament/expanse in Hebrew is "arch in the sky." Strong's Concordance #7549 defines *raqiya* as an expanse, that is, as the firmament or (apparently) visible arch of the sky. The continuous arch of water (ice) surrounding the atmosphere would be suspended in space by its own support system and other contributing factors countering the draw of gravity. The root word of *raqiya* is *raqa*, which means to expand (by hammering), to make broad, or to stretch. The idea behind the words *raqiya/expanse/firmament/heaven* is that God violently stretched out space much like a blacksmith would violently stretch out metal, not that the sky is metal, although there is metal out there, but it's how God did it with sparks flying from the violence of the rapid expansion. One of the definitions of *raqiya* is also a visible arch in the sky. The arch was a thickening of the canopy of water swelling at the equator from earth's spin and moon's gravity. This would create the appearance of an arch in the sky, and our Milky Way Galaxy, and it is why the Holy Spirit had Moses use this Hebrew word *raqiya*.

When God created the atmosphere, He created it in the "midst of the waters." This means that it was in a middle position, in the center—here, by water. The New Kings James Version uses *firmament*, and the New American Standard Bible uses *expanse* for *raqiya*; they are synonyms. The physics behind God creating the atmosphere involves the understanding that when water is pressurized and heated from the pressure and then quickly released in pressure, the water instantly boils. This is why divers are told to come up to the surface of the water slowly, so their blood doesn't literally boil, not from being hot, but from a sudden change in pressure. Therefore, when God violently and quickly stretched out the universe (Big Bang), the waters that surrounded the forming Earth instantly boiled from the change in pressure and caused an opening in the midst of the water—the atmosphere.

Since the *expanse* or *firmament* is called heaven, God said, "Let the waters below the heavens be gathered into one place and let the dry land appear." Then on the fourth day God said, "Let there be lights in the expanse of the heavens." There were two expanses created on the second day of creation. This explains why *expanse* is called *heaven* in verse 8 but pluralized when referring to the heavens one verse later in verse 9. We would only pluralize something that includes more than one. For example, "I created an Italian pie. I called the Italian pie pizza. Let the pizzas be served to the team." See how the singularity of pizza was used as the name of the Italian pie, but the plural of pizzas denoted more than one.

The canopy initially hovered spherically around the earth. How do we know this? Since our atmosphere is spherical, then the water around the sphere would initially also be spherical (like a ball).

Review: One of the definitions of *expanse (raqiya)*, also means "an arch in the sky," which matches the frozen support structure of the canopy. The atmosphere was created in the midst (middle) of waters, thus the canopy was initially spherical around the atmosphere on the second day of creation. There are three heavens: the atmosphere, universe, and the Throne of God.

We already discussed one major benefit of the canopy in Chapter 1, and that was a reduction in the net gravitational force by buoyancy force. Another benefit to life on earth from the existence of the canopy of water would have been shielding from harmful high energy from space. Light or energy comes in many forms, but higher frequency ultraviolet rays, X-rays, cosmic rays, and gamma rays are not visible to us and are harmful to life on earth. We apply lotions with various degrees of protection to protect us from ultraviolet rays, and we put lead on X-ray technicians to protect them from X-Rays. We do this because high-frequency energy damages life by causing mutations to DNA that adversely affect genes and destroy cells. Ultraviolet light breaks down collagen that connects tissue together and causes wrinkling. Additionally, X-rays, cosmic rays, and gamma rays can cause cancer. The canopy protected life on earth from the harmful effects of high-energy light penetrating down to the surface and from DNA mutations that lead to cancer or cell damage.

Before the Flood, each successive generation after Adam and Eve was a genetically closer copy than each successive generation after the Flood. The canopy of water shielded earth and significantly reduced the number of mutated genes from those harmful rays.

Some mutations occur in genes as a result of harmful radiation from the sun. The information in a DNA strand has a built-in mechanism of repair that acts by deleting the mutated portion of a gene or by fixing the mutated portion via protein markers to produce an accurate-as-possible copy of the genetic information. However, mutated DNA still occurs and may result in impaired function, cell death, or death of the host via cancer, but it never results in new information that creates a new function that does not already exist in the DNA code. It also will not result in a new kind of creature. This is the difference between adaptation and evolution. Evolutionists sometimes suggest that high energy can cause mutations in the genetic code, and the mutated DNA can result in a new code that produces a new function that was never before encoded in the DNA code. This has never been observed or tested to be valid. Some evolutionists believe that when mutated genes get passed down, additional changes occur that get passed down, and different kinds of creatures are eventually created from new information never before encoded in the DNA code.

Evolutionists believe that all living things are genetically related through one common ancestor and through a process of descent through modification. The central mechanism of evolution occurs in three steps:

    1. Mutation of DNA by external stimuli that enhance the code to eventually produce a new DNA code for a new function or new species, which eventually leads to a new kind of creature.
    2. Altered embryological development.
    3. An environmental scenario that causes the natural selection of a new type of creature.

However, mutated DNA or mutated cells or mutated genes from high-energy or even chemical stimuli resulting in mutations always has an adverse outcome and never has a beneficial effect on the DNA code for the offspring or parent. On the other hand, adaptation is the ability to modify features that better fit the environment from information already existing in the DNA code. Therefore, when a gene is mutated from high-energy rays and then based on information already existing in the DNA code, protein markers are dispatched to try and repair the mutated genetic code, not rewrite the code to try to evolve to the situation. Thus, adaptation is based on a preexisting DNA code, not a new DNA code. Therefore, evolutionists are in error to use adaptation as evidence for evolution. In addition, natural selection is also based on a preexisting DNA code, so a mutated DNA code gets eliminated, not passed on. Arguing that natural selection supports evolution is getting the cart before the horse because the information to select must be created first.

The problem for evolution is that when mutation of the DNA code occurs from catalysts such as high-energy radiation, then the host is weaker, has a distorted external appearance, becomes sterile, is stillborn, dies prematurely from predators or climate, or has impaired function, though it is still the same kind of creature. But in every situation, the opposite sex has DNA information that forces them to only select individuals who are the strongest and the most genetically accurate copies of their kind. Different kinds of creatures cannot mate outside of their kind (e.g., all cats stay within the cat family), and secondly, their DNA is preprogrammed to avoid the mutated and the weak.

In the case of asexual reproduction, when a gene code gets mutated from the sun's radiation, there is usually impaired function or a loss of function and sometimes the death of cells before the code gets anywhere near forming a wholly new meaningful alternate genetic code for a new function. For example, let's take a simple single-celled organism and say a simple genetic code for one protein is represented by "All cows eat grass." Now mind you, this is ultra simple because a genetic code that forms just one single protein is 1,500 nucleotides long. For this example, each character is a nucleotide. A single cell has around 250–500 proteins, with each having its own associated gene, so this is ultra simple. Now let's add some harmful radiation that alters the code with a series of mutations: "All cows eat grass" becomes "All *j*ows eat grass," then "All *j*ows *x*at grass," and then "All *j*ows *x*at *f*rass." Now it's to the point where the function of the message is lost, yet we are far from another functional phrase. Thus, with a loss of function, the cell dies way before a meaningful new function comes along. And this is where the evolutionary hypothesis needs too great of a miracle. Evolution not only needs a miracle once for this one genetic code to alter the function of one protein, but it needs trillions upon trillions upon trillions of successful miracles to evolve one single cell into the diversity and complexities of life that we see today. And no amount of time can solve this problem because once the single cell loses the function of the protein, the cell may die, and then the process of continually degrading the genetic information—hoping beyond hope to reach a different level of useful genetic information for a new protein function—is stopped.

Evolution consistently uses adaptation as their proof text, but adaptation is not evolution; adaptation is based on information already in the DNA code. For example, sheep adapt to sequentially cold winters by having woollier coats, but this ability to adapt to have a woollier coat is based on information already existing in the DNA code, not new information that is a result of multiple winters.

The canopy of water blocked solar ultraviolet light, X-rays, cosmic rays, and gamma rays from penetrating to the surface of the earth and thereby protected the DNA code of all life from mutation errors that can adversely affect life on earth. This allowed life to function on a cellular and genetic level at optimal design, free from the impairment or loss of function associated with a mutated DNA code, and it allowed copies of genes to be closer to those of the original Adam and Eve, which prolonged the longevity of life. In fact, scientists who are planning for humans to travel vast distances in space and without the protection of the earth's magnetic field have had to come up with a solution to protect astronauts from cosmic rays, gamma rays, and X-rays. Their solution is that the water needed for space travel and for life to be sustained will be stored around the ship's hull to block high energy from penetrating into the ship and to protect the passengers. It's the same principle and the same outcome as the canopy of water. If you want a simple explanation of waters' ability to shield an object from the higher energy of a star, just spend the summer in a pool: the parts of your body below the water line will have less tan the farther away they are from the surface.

With the existence of the canopy, there would be virtually zero or trace amounts of Carbon-14 generated from stellar energy in the atmosphere. Thus rendering C-14 dating methods today meaningless to determine the age of something that died while that canopy existed.

Review: The canopy of water shielded life on earth from harmful high-energy rays. This protected the DNA code from mutated errors and allowed life to thrive on a genetic level, with accurate copies of DNA information being passed on to offspring.

You may be wondering: if there was a canopy spherically hovering around the atmosphere, then how did life on earth get sunlight for photosynthesis? This canopy would not block the sunlight; it would allow the sun's light to penetrate through, but the harmful rays would be shielded from fully passing through—just like light penetrates water and lights up the bottom of a pool, so too would light penetrate this canopy and light up the earth. If someone stands in water, sunlight and the sun's rays still penetrate the water, but the thicker/deeper the water, the less sunlight and the fewer sun rays penetrate. The sunlight penetrates more freely with less of a reduction in strength from the thickness of water than higher energy rays (ultraviolet). For example, if an individual stood in a pool up to his or her shoulders for the summer, his or her toes would get sunlight, but they would not get enough UV rays for tanning. The canopy of water would shield life on earth from UV rays, and a portion of the UV rays would be reflected back into space.

Light penetrates through water and ice, but it is refracted or bent, which means that the angle at which the light enters water and ice is different than the angle that light exits water and ice. For example, if you were a spear fisherman, and you wanted to hit a fish near the head, you would aim for the tail of the fish because the view is changed from the water refracting/bending the light. This refraction aspect of water means that the sun's light and warmth would also be scattered abroad the earth, hence illustrating the ambient temperatures across the globe. The poles would receive enough light and warmth to prevent freezing, and the equator would have enough light and heat refracted to prevent deserts. Without ice at the poles and deserts at the equator, there would not be a global wind (jet stream).

The ice formed as the canopy in the cold of space would be colorless; there would be no soil to block light, no sediment, and no particulates to impede the transparency of light passing through. Only alkaline elements would be diluted in a clear brine, as in the South Pacific, so the ice and water would be clear. Snow is white because it reflects all the colors, and ice is clear because it transmits all the colors. They are different, so the canopy would not be white like snow, but clear.

The shape of this canopy of water was spherical at its formation on the second day of creation, but with the earth's spin and subsequent canopy's spin, centripetal force morphed the canopy toward a more disk-like shape. Shortly after the formation of the canopy, as temperatures in space began to cool after the Big Bangs, then a frozen external structure formed, probably by the end of the second day of creation. This prevented Earth's gravity from pulling the canopy down to the planet surface. With the moon's gravity causing two daily tides, the canopy would bulge nearest the moon, and the liquid from within the canopy would fill those cracks. The process of centripetal force from the spin of the canopy and the two daily tides that affected the canopy would have caused the spherical canopy to steadily morph to a bulge at the equator over decades, and then it morphed into a disk-like band around the earth over centuries. This disk-like band would be visible and look like an arch in the sky (*raqiya*). As the canopy thinned near the poles and thickened near the equator, there would be openings in the thinning canopy that would allow the atmosphere to escape to higher altitudes, thus losing a component helping to hold up the canopy, and that was the buoyancy of the atmosphere pushing up against the canopy. With the fullness of a thick disk-like formation of the canopy, the only thing holding up the canopy was its frozen arch support. Potentially, a shadow could have been cast upon the surface of Earth, but by this time, Noah was building the ark in preparation for the Flood.

An arch of ice and liquid was suspended around the atmosphere until meteors and asteroids came crashing down through the icy arch of the canopy, breaking the entire support structure, allowing gravity to bring down all that water in 40 days and nights of rain upon the earth. The inundation of meteors and asteroids coming down upon the earth could have been the means that God used to initiate the rain that led to the global Flood. And these asteroids crashing down upon the earth would have fractured the crust, released tectonic stored energy, and caused deep caverns of water hidden under the crust of the earth to burst violently into the atmosphere (Gen. 7:11).

Scientists have discovered that Saturn is not the only planet in our solar system that has rings of ice surrounding it. Uranus, Neptune, and Jupiter also have rings. The premise of the canopy is observable and testable. So as much as atheists and theistic evolutionists make fun of young earth creationists for coming to the logical conclusion of the canopy hovering around the atmosphere, there is a preponderance of Biblical support, observable support, and scientific support. Thus, the one who rejects the canopy construct is the one who is blind to logic, science, and Scripture.

Review: Even considering the water and structural ice in the canopy, light would still pass through to the earth. Harmful UV rays would be significantly reduced.

Scientists have discovered the remnants of lush vegetation that existed throughout Africa before it was converted to an arid landscape and lush vegetation in the Arctic Circle before it was converted to frozen tundra. Several ancient cultures also have stories of a great ancient city, Atlantis, that was covered up by water or some global flood story. A canopy allows for lush vegetation at both poles and no deserts. Prior to the Flood, all the water that rained on the earth for 40 days and 40 nights came from two locations. Some water came from up above the atmosphere, and some freshwater burst out of the earth from deep caverns just below the surface of the earth (Gen. 7:11). The water above the atmosphere would have created a buoyancy effect on all objects on earth, and the water below in the deep caverns would have raised the water table closer to the surface. Oceans are vastly larger than seas, so it is safe to suggest that there were only seas prior to the Flood. Prior to the Flood, vast amounts of water were above our atmosphere and in deep caverns just below the surface. Therefore, more land was covered with vegetation rather than covered with oceans.

Prior to the Flood, there were seas; during the Flood, the entire Earth was covered by water. After the Flood, the waters receded because of the glacial age, and the mountains rose and the valleys sank (Psalm 104:8), and oceans came into existence. But the ocean depths were a couple of hundred feet below their current depths for a couple of hundred years, as the glacial age melted away and raised the oceans to their current levels.

Chapter Summary: The canopy of salt water above the atmosphere would create buoyancy, countering earth's gravity and weakening gravity's strength. Initially, the canopy spherically surrounded the atmosphere; it morphed toward a disc shape because of tidal forces and centripetal force over time and was suspended above the earth by:

1. Its physical distance from the earth's full gravitational force.
2. The earth's rotational spin, which created centripetal force countering gravity's pull.
3. The atmosphere's buoyancy force pushing up on the canopy.
4. A frozen structure, which created a suspension arch in the sky.
5. Covalent bonds, which held it together.

<u>Group Questions:</u>

1. How does it change your view of the canopy concept, knowing that Saturn, Uranus, Neptune, and Jupiter (50% of the planets in our solar system) have some form of a ring system around them?

2. Which benefit from the canopy on life on Earth impressed you the most? Was it decreased gravity, increased oxygen, ambient global temperatures to prevent polar ice, deserts, tornadoes, and hurricanes, or atmospheric pressure to cause the mist, shielding life's DNA from high-energy cosmic rays that mutate the DNA, or the beauty of seeing a sparkling arch in the sky?

3. How does the theory of the canopy influence your interpretation of the age of the earth?

Chapter 4
Climate

What would a canopy of water hovering our atmosphere do to our climate? This would create a greenhouse effect, not in the negative sense as in too much CO2, but in a positive sense as in stable ambient global temperatures with high O2 concentrations. Ambient temperatures on the surface of the earth would mean no polar ice caps, no deserts, and abundant lush green tropical vegetation uniformly growing around the globe. Before the global flood of Gen. 7, there would have been massive amounts of water stored above the atmosphere in the canopy and stored below the crust of the earth in deep caverns; oceans wouldn't exist, only smaller seas. Let's journey through these claims, and see if there is support for this pre-Flood utopia.

Does the Bible support ambient temperatures before the Flood? Yes. Gen. 3:7: "Then the eyes of both of them (Adam and Eve) were opened, and they knew that they were naked; and they sewed fig leaves together and made themselves loin coverings." That is correct—Adam and Eve lived in a climate that was so ambient that they were naked every day and night of the year until they sinned and covered only their private parts. Notice that they didn't put on jackets for winter or clothes after they sinned. Therefore, the climate didn't change. Their eyes were spiritually open to the knowledge of good and evil. With a canopy sealing in the earth, then the earth itself would also warm the atmosphere and produce ideal ambient temperatures. One benefit: there would be no wasted energy to warm a cold body, no shivers and goose bumps, and less panting or sweating to cool an overheated body.

Do evolutionary scientists support the notion that at some point, there were ambient global temperatures? Yes, according to their belief system, they contend that billions of years ago, conditions were just right for a primordial pool of complex chemicals to spontaneously spawn life. In addition, they further believe that during the Cambrian explosion of life, oxygen levels were high enough and temperatures stable enough for life to thrive and eventually lead to a great variety and complexity of creatures; there was a brief (in geological terms) period of optimal temperatures that allowed life to thrive. Optimal conditions for life is a foundation for evolution. Though we differ on when, what, how, and who, both evolutionists and creationists believe there were periods of optimal conditions for life.

Does the Bible support the scenario of no polar ice caps and no deserts before the Flood? Yes. The existence of polar ice caps and deserts creates global changes in temperature from one region to another on the earth. These differences in temperature cause a global wind called the jet stream. When did the wind first occur? Immediately after the Flood of Gen. 8:1: "God caused a wind to pass over the earth." A likely explanation of how God did this is that the canopy reduced the sun's harmful heat on the equator and dispersed essential heat to the polar regions. This is corroborated by the discovery that tropical forests once covered the polar regions and why archeologists are finding camels, saber-toothed tigers, and so forth, and tropical forest below the ice in the arctic circle.

With high atmospheric pressure, no clouds can form, and thus no rain. As the canopy of water hovered over the atmosphere, this caused a great deal of weight and pressure. Therefore, there was not one single cloud on Earth from creation until the Flood. Then, with a canopy hovering around the atmosphere, when evaporation occurred each day, what would happen to the moisture? This moisture would not and could not have formed clouds because of the high pressure created from the canopy. Meteorologists explain that with high pressure over a region, there are no clouds. And with a low pressure system over a region, there may be a thunderstorm. So what would happen to evaporated water? During the cool of the day, the moisture in the air would rest on the soil as dew. Gen. 2:5–6 "For the Lord God had not sent rain upon the earth, and there was no man to cultivate the ground. But a mist used to rise from the earth and water the whole surface of the ground." With a mist rising from the ground to water the earth, no extreme equatorial heat, and no polar ice caps, then there were no large changes in global temperatures. Therefore, there were no tornadoes, no hurricanes, no floods, no droughts, no lightning, no snow, and no hail before the Flood. With the loss of the canopy and the subsequent reduction in atmospheric pressure, the addition of extreme heat over the equator because the canopy no longer shielded it from the sun, and with the polar ice caps formed because of the Flood, then the new means for watering the ground—clouds, rain, and rainbows—also led to hurricanes, tornadoes, hail, lightning, floods, droughts, and so forth. These weather events are a reminder of the curse of sin placed upon all creation. The loss of the canopy caused the global flood and reduced atmosphere pressure, and this ended the mist rising from the ground to water the whole earth.

As a result of losing the canopy, then came the change of seasons from cool springs to hot summers and from the hot summers to the cooler falls and winters. This resulted in destructive and harmful tornadoes and hurricanes, which didn't occur before the Flood because of the relatively constant ambient global temperatures.

The first mention of winter, summer, cold, and heat is not until after the Flood in Gen. 8:22. Let that sink in for a moment. This is strong evidence that before the Flood, there were no winters, no hot summer, no cold, and no hot days of summer. In Gen. 1:14, when God finishes forming the sun, moon, and stars and says, "Let them be for signs and for seasons." This Hebrew word for seasons doesn't mean spring, summer, fall, and winter. No—it means feasts and festivals and appointments of assembly. This is a common mistake by casual or surface readers of the Bible.

Review: From creation till the Flood, there were mild, ambient global temperatures, and this helped life to thrive. There were no ice caps, no deserts, and no global jet stream, only mild breezes. The hydrological cycle bypassed clouds and rain and converted evaporation into a mist because of atmospheric pressure.

Romans 8:18–23 sheds more light on the climate at the time of Adam and Eve and the dinosaurs. Verse 20 reads, "The creation was subjected to futility, not willingly, but because of Him who subjected it." This means that as a result of Adam and Eve's sin, all of creation was subjected to the same judgment from God, not just humans. All of creation needs a savior to save it from the curse of sin. Currently, our climate goes through a symbolic manifestation of the effect of sin, the four seasons: death (fall), burial (winter), resurrection (spring), and life (summer). Romans 8:21: "In hope that the creation itself also will be set free from its slavery to corruption into the freedom of the glory of the children of God. For we know that the whole creation groans and suffers the pains of childbirth together until now." This means that as the children of God will receive an immortal body free from death, so too will all of creation return to a "pre-sin" state and no longer suffer death (fall), burial (winter), and the suffering of a decaying body (hot summer). Therefore, before the fall of Adam and Eve, creation was not subjected to futility and was not "groaning to be reborn," and therefore, there were no four seasons, no death (fall), no burial (winter), and no hot summers. That is why Adam and Eve were naked; they lived in perfect temperatures every day and every night because the canopy around the atmosphere created perfect global temperatures. The effects of Adam and Eve's sin was not fully manifested until after the Flood. Then came the four seasons, adding hot summer, fall, and winter to the preexisting spring and mild summer.

Also, prior to the sin of Adam and Eve, there were no active volcanoes, no earthquakes, no droughts, no tsunamis, and so on. There were no natural disasters because there was no "groaning and suffering" of creation to be set free from the corruption of sin (Romans 8:18–23). The corruption of sin leads to death, physically and spiritually. With "a mist used to rise from the earth and water the whole surface of the ground" (Gen. 2:5–6), this process of providing the earth with a mist was free from all natural disasters associated with clouds, rain, and the global wind, such as floods, droughts, hail, tornadoes, hurricanes, and so forth. In addition, toward the end of the Flood sage, God caused the valleys to sink down and the mountains to rise to the heights we see today (Psalm 104:6–9). It is the mountain ranges that also prevent clouds carrying moisture to pass over them and water the parched land on the other side. Thus, prior to the Flood, while the mist rose from the ground, there were no tall mountains to cause droughts, only smaller hills and smaller mountains.

Scientists discovered that the earth was once covered with a supercontinent called Pangaea. This is in accord with the Bible because when most of the ocean waters were stored up in the canopy and stored down below in the deep caverns, then the surface of the earth was potentially 70% usable land for vegetation—a supercontinent. Thus, before Pangaea broke up into the continents we see today, there were far fewer earthquakes. From a Biblical perspective, there were no earthquakes before sin. And while Pangaea remained as one land mass, the tectonic potential energy mounted and tension built up until the Flood when Pangaea broke apart. Earthquakes represent an unstable Earth adjusting itself to release tension and pressure from one tectonic plate moving against another tectonic plate. This movement and release of tension indicates a fallen state and a state of unease. This doesn't sound like the perfect creation that God describes in Gen. 1; no, this sounds like a fallen state of unease—sin. Therefore, it is hypothesized that there were no tectonic plate movements (earthquakes) before sin occurred on earth, I could see God causing a mild tremor with the first sin, but Pangaea didn't break apart until the Flood. Therefore, a hypothesis of this book is that potential energy was building up from the first sin until the Flood (~1,650 years worth), which is when Pangaea broke apart, and all that potential energy was released as kinetic energy.

Tectonic plate movements often allow cracks in the crust of the earth for lava to pour through, so one of the hypotheses of this book is that before tectonic plate movements and before the Flood, there were no volcanoes.

Review: From creation until the Flood, there were no four seasons, just spring and a mild summer. Before sin entered the world, there were no natural disasters, such as earthquakes, volcanoes, tsunamis, and droughts. These represent a cursed creation because of sin.

Let me illustrate the impact of a canopy and the atmosphere on the temperature of the surface in another way by discussing what happens when there is no atmosphere. The moon has massive temperature swings from the light side versus the dark side. It has a daylight high of 130°C (265°F). The dark side has a low of -110°C (-170°F). Why? It's because the atmosphere no longer exists. An atmosphere creates an environment of moderately stable temperatures. An atmosphere with a canopy around it would create an environment of even more stable temperatures. Not only did we have an atmosphere, but we had a canopy of water that further solidified stable and relatively uniform global temperatures day and night. For Adam and Eve to be comfortably naked, the nighttime lows were probably around ~75°F, and the daytime highs were probably around ~85°F.

Does the Bible support lush vegetation globally? Yes. With no wind prior to the Flood, there were no polar ice caps and no deserts to drive the jet stream and no oceans, only seas. If you think having no oceans is an impossibility, consider Rev. 21:1, "Then I saw a new heaven and a new earth; for the first heaven and the first earth passed away, and <u>there is no longer any sea</u>." Not only was there no ocean, but there were no seas as well. Gen. 1:10: "God called the dry land earth and the gathering of the waters He called seas." Where was all the ocean water then while the earth only had seas? Since the global flood caused the oceans, and the sources of the Flood were the canopy and the deep caverns, then potentially before the global flood, 33–50% was stored in the canopy, 25–33% was stored in the deep caverns under the crust, and 25–33% was the seas, which leaves more dry land for vegetation growth than what we see today because there were no oceans, no polar ice caps, and no deserts. That dry land was covered in lush vegetation, and a mist/flow of water rose from the ground to water the whole earth. Not much grows in deserts or at the polar caps; it's just not good for life. When God made all the plants/trees/grass on the third day, "God saw that it was good" (Gen. 1:12)—not barren as represented by deserts and polar ice caps, but good as in lush, abundant, and thriving. This is why remnant tropical forest are found below the ice of the polar regions and the Saharan desert.

Review: Today the oceans and polar ice caps and deserts limit or prevent vegetation growth that produces oxygen for life to thrive. Before the Flood, there were no oceans, no ice caps, and no deserts, allowing massive amounts of vegetation to grow and an increased production of oxygen, which allowed life to thrive.

The canopy created a high buoyant force, coupled with the caverns of the deep, that allowed water to evaporate up as a mist and then settle on the earth as dew (Gen. 2:5-6). When did the mist rising from the ground watering the earth stop and the clouds and rain cycle take over? This was at the time of the global flood. Gen. 7:11: "All the fountains of the great deep burst open," and the canopy rained on the earth for 40 days. Resulting in a reduction of the atmospheric pressure, and buoyant force, which increased gravity, which stopped the mist rising from the ground and allowed clouds to form for rain. Like a chain reaction, each change effecting another. Then, God started the global wind—the jet stream. Gen. 8:1: "God caused a wind to pass over the earth, and the water subsided." For global wind to occur, there needed to be large changes in global temperatures. For the hydrological cycle to occur with cloud formation and rain, there needed to be higher equatorial heat and a low-pressure atmosphere to allow cloud formation and rain. This doesn't mean low humidity before the Flood, there was probably high humidity just like tropical forest today, but the high atmospheric pressure from the canopy prevented cloud formation. The process of rain falling under direct sunlight also led to a new phenomenon of light refraction from water droplets, creating rainbows. Gen. 9:12–14:

> This is the sign of the covenant which I am making between Me and you and every living creature that is with you, for all successive generations; I set My bow in the cloud, and it shall be far a sign of a covenant between Me and the earth. It shall come about, when I bring a cloud over the earth, that the bow will be seen in the cloud.

Therefore, prior to the Flood, because of the higher atmospheric pressure from the canopy, there were no clouds, no large temperature changes to cause wind, no rain, and no rainbows. I imagine there were gentle global breezes because of nighttime temperature drops and daytime temperature increases, but not wind (i.e., no jet stream).

Review: Before the Flood, there were no rainbows, no rain, and no clouds. The atmosphere was denser, humidity was probably higher, and a mist rose from the ground to water the earth.

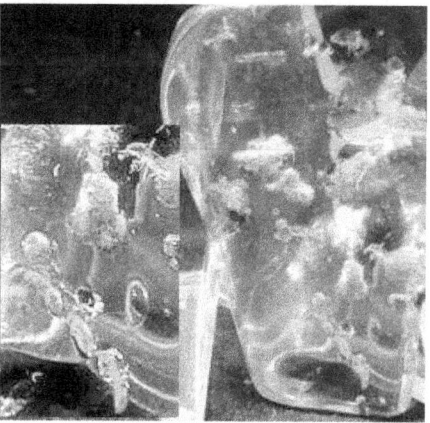

What would the oxygen concentration of our atmosphere be with the earth containing global tropical forests, with no oceans but only seas, no deserts, no polar ice caps but lush forests and vegetation? There would probably be a 50% increase in oxygen concentration, and that is what scientists have discovered. The current oxygen concentration in our atmosphere is 21%, and when dinosaurs roamed the earth, the O_2 concentration was ~31%. Higher oxygen concentrations are exactly what have been found in amber bubbles and deep air bubbles in the polar ice caps. The canopy was the key to creating an environment with higher concentrations of oxygen. With no polar ice caps, no deserts, no oceans, only seas, this means that there was much more land for massive amounts of lush tropical vegetation encompassing the earth. More vegetation equals more oxygen.

That oxygen-rich environment would have caused life to thrive, with accelerated growth and larger life forms. This oxygen-rich atmosphere explains how some dinosaurs lived with small nostrils and small lungs compared to body mass. Brachiosaurus had nostrils the size of a modern-day horse. Their survival and thriving in life was possible due to the high oxygen concentrations and higher pressure before the Flood. If brachiosaurus lived today, their nostrils and lungs would be too small to allow enough oxygen in to support their large size, and they would physically suffer or die. The oxygen-enriched atmosphere allows us to understand how dinosaurs flourished and grew so large and why humans lived so long in the past—because oxygen causes life to thrive. Given that our atmosphere has less oxygen now than when the dinosaurs roamed the earth, this sheds light on subsequent genetic adaptations to changes in oxygen concentration and gravity intensity. Now, we need to discern when or how this oxygen-rich atmosphere changed. The Bible gives us the answer: the Flood. This flood was global, and every mountain was covered (Gen. 7:19–21). It rained for 40 days and 40 nights. And the water stayed and covered the earth for 150 days (Gen. 7:24). This would have resulted in massive polar ice caps and created an ice age.

How did dinosaurs and humans coexist in the Book of Job (Chapters 40–41) after the Flood, yet not coexist today? Two major changes on the earth solve this mystery: the gradual reduction in oxygen concentration in the atmosphere and increased gravity. These two results have altered the living conditions on earth, and life adapted because of the information embedded in the DNA code from creation, and natural selection—based on those DNA codes—has allowed certain species to dominate populations in certain regions of the globe. This means that living beings don't grow as tall or live as long as they once did before the Flood. Dinosaurs could still be living amongst us, but they are just too small for us to recognize them as dinosaurs, so we call them by other reptilian names.

With the canopy supported in space by a frozen arch and a central reservoir of water to fill any cracks, spinning with the earth and remaining in a geosynchronous orbit, something needed to mechanically break the arch support to cause this canopy to be pulled into the earth's atmosphere and result in rain for 40 days and nights. Something did happen; the earth has the scars to prove it, and so does the moon. Potentially, the answer is asteroids. Multiple asteroids and meteors hitting the frozen arch support structure of the canopy, along with the force of gravity, would have caused the canopy to slowly come down upon the earth as rain. A series of space objects hitting earth would affect life on earth and would cause massive fires. It would also cause soil and particulates to commingle with the air and fractures in Pangaea, which would set in motion the tectonic plates and volcanoes. A "dinosaur killing" impact at the Yucatan Peninsula is the likely culprit.

The Yucatan Peninsula crater is called Chicxulub. It is a massive impact zone and one of the larger impact craters on the earth. There are hundreds of large meteor and asteroid impact craters on the earth, but Chicxulub was potentially the major occurrence that initiated the Flood upon Earth. This is because it landed near the equator, which means that it passed through the space that the canopy would have been occupying. There is certainly enough evidence in the soil showing a layer of ash from asteroid impacts in the past. That thin layer of ash from the fires caused by the asteroids is covered by many layers of soil from the muddy flood waters. We know that there was a period of heavy asteroid bombardments that formed a layer of ash because of iridium present in that thin layer of ash. Iridium is not naturally found on earth. The moon also shows the same scars from the same time in history, when both the moon and the earth were bombarded by asteroids and meteors.

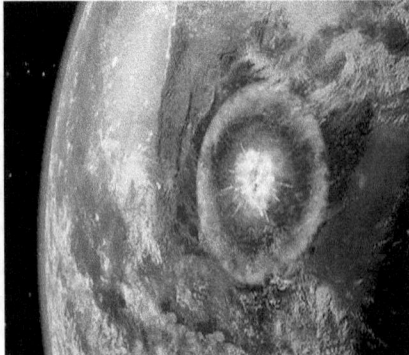

Geologists confirm that a dinosaur killer asteroid impacted the earth in the past, although evolutionary geologists hypothesize that this occurred hundreds of millions of years ago. One evolutionist said that if an asteroid caused the Chicxulub crater, it would have caused a massive tsunami. But this notion is based on the idea that the oceans existed before the Flood. However, the layer of ash with iridium wouldn't be distinct and clear as it would be if the asteroid impacted an ocean. A large portion of the water in the oceans today was stored up above the atmosphere in the canopy, and another large portion was stored below the surface. The remaining portion of ocean water is from the seas that were on earth from the time of creation till the Flood. Therefore, when asteroids hit the earth at such places as the Yucatan Peninsula, they probably hit land and caused the land to fracture and broke Pangaea like an egg. *Picture Credit: Everything Dinosaur.* http://blog.everythingdinosaur.co.uk/.

Chapter Review: The climate before the Flood had ambient global temperatures, no oceans but seas, more land exposed that was covered with vegetation, no polar ice caps, no deserts, higher oxygen concentration, higher buoyancy force, weaker gravity, no global jet stream, no hydrological cycle, and no rainbows. Potentially, an arch formed from the frozen canopy of water and was visible to those on Earth. Additionally, living beings inhaled more oxygen with each breath, all vegetation was larger, dinosaurs roamed the earth, mankind lived 900+ years, nights were not as cold, and days were not as hot—a perfect Utopia for optimal growth and life.

Group Questions:

1. With Adam and Eve living without clothes, how does that influence your understanding of temperatures during their time?

2. Knowing that there were tropical forests in the polar regions, how does that shape your view of global temperatures before the Flood that covered them in ice?

Chapter 5
Oxygen Concentration

Oxygen concentration in our atmosphere is vital to life. Without oxygen, life on earth would cease to exist. Currently, the oxygen concentration in our atmosphere is around 21%, and 79% of the molecules in our atmosphere are from other elements (78% nitrogen and <1% other). Since a lack of oxygen kills life, and a reduction in oxygen causes life to suffer, would an increase in oxygen make life thrive? What would happen to life on earth if the oxygen concentration in the atmosphere was 50% higher than today? This would mean that oxygen in the atmosphere would make up ±31% of the atmosphere.

A research team at Arizona State University did a study of insects in an atmosphere at 31% oxygen concentration (a hyperoxia environment), which is a 50% increase in our current oxygen concentration in the atmosphere. The researchers also studied insects in an atmosphere at 12% oxygen concentration (a hypoxia environment).

The results were staggering; most of the insects raised in hyperoxia conditions were larger in size than normal, which clearly demonstrates a thriving life. Only the cockroach remained the same size. However, the cockroach's internal tracheae (tubes that carry oxygen throughout the body) were smaller than normal as a result of living in a 31% oxygen environment, which allowed other organs and tissues to grow larger, such as muscles, blood vessels, tendons, ligaments, and so forth, which means those cockroaches would be faster, stronger, and quicker and have improved endurance. Therefore, all insects in an oxygen-enriched environment would thrive. In a hyperoxia environment, insects grow larger. And the ones that don't seem to grow larger (because an exoskeleton restricts growth) would eventually grow larger in time as they adapted to the increased oxygen concentrations because of the ability to adapt already exists in the DNA. This is called adaptation or "speciealization," this is not evolution because they are still the same kind of creature.

Conversely, the research showed that the insects grown in hypoxia (low oxygen concentration, around 12% oxygen concentration, which is half of today's value) were smaller in size. And even though the cockroach remained the same size, its internal tracheae (tubes that carry oxygen throughout the body) grew larger to compensate, which forced other organs and tissues to be smaller. Having smaller tissues, such as those related to muscles, tendons, blood vessels, and nerves, means that those cockroaches raised in hypoxia (low oxygen concentration) would be less active and unable to run as long or as fast or jump as high. This is a negative approach to further prove that higher oxygen concentrations cause life to thrive.

Review: Experiments show that higher oxygen concentration in the atmosphere equals thriving life. Life will grow larger, faster, and with greater longevity in an environment with higher oxygen concentrations.

Today, we have cockroaches that are about 2.75 inches long, but the fossil record shows cockroaches that grew to 18 inches long. That is 6.5 times larger. Today, we have dragonflies with a wing span of 6 inches, but the fossil record shows dragonflies with wingspans up to 50 inches. That is 8.3 times larger. Today, the chambered nautilus (it looks like a squid in a shell) grows to about 10 inches in diameter, but the fossil record shows that they once grew to 8 feet in diameter. That is 9.6 times larger.

How did scientists determine that the earth's atmosphere had a 50% higher oxygen concentration in prior times versus today's atmosphere? They bored into amber bubbles and tested the oxygen concentration inside the trapped bubbles. In some past millennium, when tree sap oozed out of a tree, the sap became trapped air in bubbles. Then the sap petrified as amber with air trapped inside from prehistoric times. It should be noted that sap will only petrify because of a sudden covering of a lot of sediment, moisture, pressure, and heat. Nothing petrifies or fossilizes with a slow covering of sediment and moisture over 100,000 years.

Another way that mankind has determined that oxygen concentration was higher in past millennia is through deep glacier drilling and testing of the air bubbles trapped inside those deep glacier core samples. When core samples were examined, researchers noticed that there were pockets of air bubbles. They tested the captured air pockets and determined that oxygen levels were 50% higher in the distant past.

Review: Scientists have discovered that in the past, the earth's atmosphere had 50% higher oxygen concentration levels than today. This was discerned via air bubbles trapped in amber and glacial ice caps.

There are several examples of how a low oxygen concentration can starve life, reduce life's size, decrease longevity, and kill life. There are a whole host of diseases and ailments that are associated with a lack of oxygen in the body, from paleness, lethargy, ischemic necrosis, avascular necrosis, anemia, gangrene, sickle cell anemia, brain damage, heart attack, and death. There are many names given to conditions and diseases resulting from a lack of oxygen to the body or that cause a lack of oxygen to specific organs and tissues in the body. Why? Because God breathed life into all creatures and humans (Gen. 2:7 and Gen. 1:30). Take away the blood that carries the oxygen or take away the oxygen, and you take away the life. Therefore, when someone's blood is lost, they will die, or when someone's breath is taken away, they will die. Therefore, life is in the blood and breath (Lev. 17:10–14). Both deal directly or indirectly with getting oxygen to the body.

No one disputes the importance of oxygen. And research clearly proves that life thrives with higher oxygen concentrations and suffers with lower oxygen concentrations. But what does this have to do with *Origins*? Well, one of the major tenants of this book is that oxygen concentration was much higher in past millennia than today. And this higher oxygen concentration in the past allowed animals and plants to grow to enormous height and weight and to live a much longer life. Dinosaurs stood 100 feet tall and weighed 100 tons. Also, humans lived 900+ years.

Review: There are many diseases and ailments that occur due to a lack of oxygen to a particular body part or to the entire body in general. Simply put, a lack of oxygen kills or decreases life's potential. Also, God teaches that life is in the breath and in the blood. Both directly and indirectly get oxygen to the body.

If we could discern how and when the oxygen concentration dropped from 31% to 21%, that would tell us approximately when dinosaurs stopped growing so large and when humans stopped living 900+ years. If we look at what is preventing oxygen from being at a higher concentrations today in the atmosphere, maybe we can work backward to determine what caused oxygen levels to decline and why, how, and when. To have oxygen in the atmosphere, we need vegetation. Plants, grass, and trees produce oxygen. To determine what is preventing oxygen in our atmosphere from being at 31% concentration as in prehistoric times and keeping oxygen at 21% concentration today, we need to determine what is preventing vegetation growth.

The primary limiting factor that physically limits vegetation growth is the ocean. The large oceans cover up vast amounts of land and thereby prevent oxygen-producing vegetation on the land from growing. Currently, water covers around 70% of the earth's surface.

The second limiting factor that has reduced the oxygen concentration in our atmosphere and has prevented oxygen concentration from rising is the lack of uniform temperatures around the globe. The variance of global temperatures around the globe, from hot at the equator to cold at the poles, has created two results: hot deserts and cold polar ice caps. These two regions physically prevent oxygen-producing vegetation from growing on 10% of the earth's surface. The arid, dry deserts are too hot and lack enough water to allow vegetation to grow. And the arid polar ice caps are too cold to allow vegetation to grow. That leaves only 20% of earth's surface for vegetation growth to produce oxygen.

Therefore, since the oceans, polar ice caps, and deserts prevent higher oxygen concentrations in our atmosphere, and since oxygen concentration was indeed higher in the distant past, then we need to determine when there were no polar ice caps, deserts, and oceans. Then we can determine when oxygen concentrations in our atmosphere were higher. And by doing that, we'll find the cause.

Geologist Peter deMenocal of The Earth Institute, Columbia University, studied Atlantic core samples just off the coast of the Sahara Desert. He discovered that the Sahara is a recent phenomenon in geological terms and is roughly 5,000 years old. He discerned that the easterly winds tend to blow arid sand into the Atlantic ocean, and by taking core samples and counting seasons like rings of a tree, he was able to come up with an estimated age of the Sahara Desert. His research also revealed that the Sahara had lush vegetation just prior to desertfication.

This fits nicely with the Bible, as deMenocal's research has a time line of when the lush tropics of North Africa became deserts. This is near enough to when the Flood occurred, which resulted in too much heat and loss of moisture in the Sahara region, changing it from a tropical area to a desert. With the loss of lush vegetation producing O2 in the Sahara region after the Flood, O2 levels decreased.

The deserts are caused by being too close in proximity to the sun's harmful heat and lack of precipitation, and the polar ice caps are caused by being too distant from the sun's beneficial heat. Both conditions are remedied by uniform global temperatures. This means that before there were polar ice caps and deserts, the earth had uniform temperatures globally. But there is still one problem. When did the vast oceans form to cover up the vast amounts of land that produced O2? To determine when oxygen levels were higher in our atmosphere, we need to determine when the vast oceans were only seas. It's not enough to convert lush vegetation into deserts, and it is not enough to convert lush vegetation to polar ice caps. We also need to find an event that formed the oceans and subsequently formed the polar ice caps and deserts as well. The vast volumes of water in the oceans need to be stored somewhere for more land to appear and allow oxygen-producing vegetation to grow.

Therefore, the key in unlocking this mystery of "when" and "what" caused the reduction of our oxygen concentration from 31% to the present day 21% is to find a global catastrophic event that not only formed the polar ice caps and the deserts but also formed the vast oceans and changed global temperatures from uniform to regions that are very cold and other regions that are very hot. It seems like we are stuck. How are you going to get rid of trillions of gallons of water? There are only two choices: either the water came from beyond our atmosphere, and/or it came from within the earth itself, from deep caverns. There are many cultures that say there was a lost city due to a great flood. Some cultures call this lost city Atlantis. But we don't get closer to when the oceans were formed and covered up the landscape through cultural legends.

There is such a recording of a global catastrophic event in the earth's history. And this global catastrophic event is also recorded by mankind and nature. And it occurred in the not-too-distant past. The event is the global flood recorded by the Bible and by approximately 25 other cultures around the globe. The Bible gives enough details regarding this earth-changing global flood to discern that this event fits perfectly as a cause for our oceans to form, for temperatures to change, and for subsequent polar ice caps and deserts to form, resulting in the reduction of O2 production.

Before we get to the recording of the global flood, we need to determine where the water came from to cover the entire earth. Did the water come from space and/or from deep caverns within the earth? The answer is both. We turn to Gen. 1:2: "The earth was formless and void, and darkness was over the surface of the deep, and the Spirit of God was moving over the surface of the waters." Okay, before the earth was formed into a sphere, it was formless and void of life. But notice that there was water everywhere. Gen. 1:6: "Then God said, 'Let there be an expanse in the midst of the waters, and let it separate the waters from the waters.' God made the expanse, and separated the waters which were below the expanse from the waters which were above the expanse." The universe was expanded and caused the expanse of our atmosphere. The waters below the expanse later become seas, and the waters above the expanse become a canopy of salt water. This canopy is critical. For while this salt water remained above the earth's atmosphere, it provided ambient temperatures globally. This means there were relatively uniform temperatures from the equator to the poles, such that there was no global wind (jet stream). This canopy then came down to the earth's surface in a deluge of 40 days and 40 nights of constant rain.

Now, let's go back to Genesis for more clarity. Gen. 1:9: "Then God said, 'Let the waters below the heavens be gathered into one place, and let the dry land appear; God called the dry land earth and the gathering of the waters He called seas." Notice that there are no oceans, only seas. Oceans are vastly larger than seas. The Bible doesn't mention the word *ocean*, instead the Bible uses the word *deep* to refer to oceans, and thus distinguishes between the smaller *seas* and the larger oceans. For example, in Job 28:14 both the deep and the sea are mentioned in the same verse. This draws the reader's attention that they are different bodies of water. And since the deep is mentioned in Gen. 1:2 while the earth was formless and void of life, and since the earth was made out of water—the deep (2 Peter 3:5), and in Gen. 1:9 only the seas are mentioned, then a fair conclusion is that on the third day of creation God formed seas (and not oceans) on the surface from the deep that once surrounded the earth.

The Bible describes the existence of a canopy that surrounded our atmosphere. This canopy of water would have created a tropical greenhouse effect, thereby resulting in ambient temperatures around the globe. Also, this canopy of salt water explains where the water came from for a global flood of 40 days and 40 nights upon the earth.

There was another place for the storage of water, which when released, led to the formation of our vast oceans. We see this recording in Gen. 7:11: "In the six hundredth year of Noah's life ... on the same day all the fountains of the great deep burst open, and the floodgates of the sky were opened. The rain fell upon the earth for 40 days and 40 nights." That is saying that water burst out of deep caverns within the earth violently and up into the sky. Not only do we have the water coming down on the earth, but we also have great deep caverns bursting water into the sky. The combination of the two sources of water formed our oceans.

There we have it: before the Flood of Gen. 7 that caused the oceans, from Adam to Noah's time, there was a canopy of salt water surrounding our atmosphere and deep caverns below the surface. This resulted in no deserts and no polar ice caps, and since the Flood had not occurred yet, the vast oceans were stored in that canopy and in deep caverns within the earth. With ambient temperatures and no oceans, there would be significantly more vegetation on earth than what we see today. This massive amount of vegetation would produce significantly more oxygen than our current vegetation does today. Now, we have a record of when the earth had 50% higher oxygen concentration. That record is in the Bible and is evidence of how and why our atmosphere had much higher oxygen concentrations.

In addition, before the Flood of Gen. 7, the canopy held 33%–50% of the current ocean water. This created higher atmospheric pressure, and coupled with 50% higher oxygen concentrations, allowed a greater tidal volume of oxygen to enter the body with each inhalation. This means that more oxygen entered the body with each breath. This caused life to thrive and made life easier to live. In the sports world, there is an illegal technique of enhancing performance by blood doping. Blood doping is the practice of removing blood from the body a week prior to an event; the body will produce additional blood, and then the day of the sporting event, the blood that was withdrawn is put back into the body. This means that the body has extra red blood cells (RBCs) to carry extra oxygen to the muscles, which gives an unfair advantage to the athlete with the extra RBCs. That athlete will not get tired as quickly and will have greater endurance. Hence, it is illegal. Hence, extra oxygen to the body causes life to thrive.

Review: Prior to the Flood (±2,350 BC), the earth did not have oceans, no polar ice caps, and no deserts because they limit O2 production. The earth had ambient global temperatures. All this combined resulted in vast amounts of vegetation, which produced massive volumes of oxygen. And the result was that earth's atmosphere had 50% higher oxygen concentrations before the Flood of Gen. 7.

Higher oxygen concentrations in the past allowed animals and vegetation to grow to enormous height and weight and to live almost a millennia. Dinosaurs stood 100 feet tall and weighed 100 tons. The Bible records that most humans lived 900+ years before the Flood. Gen. 5:5: "So all the days that Adam lived were nine hundred and thirty years and he died." The first man, Adam, lived 930 years. His son Seth lived 912 years (Gen. 5:8). Seth's son Enosh lived 905 years (Gen. 5:11). Enosh's son Kenan lived 910 years (Gen. 5:12). Kenan's son Mahalalel lived 895 years. Mahalalel's son Jared lived 962 years (Gen. 5:20). Jared's son Enoch lived 365 years, and then he walked with God and went to heaven without dying (Gen. 5:24). Then Enoch's son Methuselah lived 969 years (Gen. 5:27). Methuselah's son Lamech, lived 777 years (Gen. 5:31). Lamech's son Noah lived 950 years (Gen. 9:28). You can see that people from all the generations before the Flood all lived long lives averaging 900+ years. The Bible gives the age of the fathers when they had their son. This is how Archbishop James Ussher came up with 4004 BC as the date when Adam was formed.

Then, with the advent of the global flood, along came massive oceans, polar ice caps, and deserts. They decreased the amount of vegetation, which reduced the production of oxygen. And as populations of all life increased, this reduced the oxygen concentration in the atmosphere and made life harder to sustain. After the Flood, in the recordings in the Bible of the genealogy of Noah, you can see that life expectancy decreased with each passing generation as oxygen levels gradually reduced. And because of the DNA code, they adapted to less and less oxygen until we reached a homeostasis of life expectancy with oxygen concentration. Noah's son Shem lived 600 years (Gen. 11:10–11). Shem's son Arpachshad lived 438 years (Gen. 11:12–13). Arpachshad's son Shelah lived 433 years (Gen. 11:14–15). Shelah's son Eber lived 464 years (Gen. 11:16–17). Eber's son Peleg lived 239 years (Gen. 11:18–19). Peleg's son Reu lived 239 years (Gen. 11:20–21). Reu's son Serug lived 230 years (Gen. 11:22–23). Serug's son Nahor lived 148 years (Gen. 11:24–25). Nahor's son Terah lived 205 years (Gen. 11:32). Terah's son Abram, who became Abraham (the father of the Jews and Muslims), lived 175 years (Gen. 25:7). Abraham's son Isaac lived 180 years (Gen. 35:28–29). Isaac's son Jacob lived 147 years (Gen. 47:28). Jacob's son Joseph lived 110 years (Gen. 50:26). You can see that with each passing generation, the effect of oxygen reduction is proportionately affecting the longevity of life. Within 10 generations, the average life expectancy declined from 900+ years of life to 175 years of life for Abraham. And only three generations after Abraham, the average life expectancy is reduced to 110 years.

Following time lines of genealogies in the Bible, scholars have determined: The date of the Genesis Flood was ±2348 BC; Abraham lived from 2166–1991 BC. The time span from the global flood to the death of Abraham was only 357 years. Therefore, in 357 years, life expectancy was reduced from 900+ years to 175 years.

Review: Before the Flood, humans lived 900+ years. After the Flood, life expectancy dropped dramatically. In 357 years (10 generations), life expectancy dropped from 900+ to 175 years.

Why is this significant? For one major reason: it's easy to see that life expectancy declined with each passing generation after the Flood. All the while, this reduction in life expectancy was paralleling the gradual decline of oxygen concentration in the atmosphere. After the Flood, the vast oceans covered

70% of the land, the polar ice caps prevented vegetation from growing and producing oxygen, and the deserts did the same; all three of them reduced the production of oxygen. While oxygen levels were decreasing in production, human life and animal life were increasing their consumption of oxygen as they increased in population.

With the research revealing a connection between the quality and quantity of life and oxygen concentration, it is easy to see that the reduction in the longevity of life and the post-Flood environment are inextricably linked.

The second major important consideration in terms of the connection of quality and quantity of life with oxygen concentration, is the book of Job. In the book of Job, Chapter 42:16 reads, "After this, Job lived 140 years, and saw his sons and his grandsons, four generations." Since Job was a fully mature adult at the beginning of the book that bares his name and Job lived long enough to father 10 children and to accumulate massive wealth, then Job was probably 60+ years of age when the book starts. This means Job lived around 200+ years in total. Because "After this" is referring to "after all the troubles written in the book of Job," he lived an additional 140 years. This puts the time line of when Job lived and had all his trials between the sixth and ninth generation after the Flood. This is noteworthy because the book of Job describes dinosaurs in Job 40:15–41:10 and 41:15–34. God illustrates His awesome power, and the most impressive creature He has created when He says to Job, "You can't even conquer what I've created, so who are you to question me?" Here are a few references to dinosaurs:

"Behold now, Behemoth [Behemoth means gigantic or colossal beast], which I made as well as you; He eats grass like an ox. Behold now, his strength in his loins and his power in the muscles of his belly. He bends his tail like a cedar." Cedar trees in the Mesopotamian area at Job's time were massive trees that were 100–120 feet tall. This creature that God describes to Job is so large that this Behemoth's tail is likened to a cedar tree that was 100–120 feet tall. What animals have tails 100 feet long? The brontosaurus fits that description, but not the hippopotamus for they have small a tail. And in chapter 41, God describes an even more impressive fire breathing dinosaur in the oceans.

Since the book of Job describes dinosaurs in Chapters 40–41 and Job lives in the sixth to ninth generation after the Flood, and since we don't see dinosaurs today, then (a) dinosaurs gradually all became extinct after the book of Job, or (b) dinosaurs are still living amongst us but too small to be called dinosaurs. Their life spans are too short (because of a lack of oxygen and the strength of gravity) now for them to grow to the same large sizes as before. Therefore, some may still live amongst us but are too small to be recognized as dinosaurs and are seen as just large reptiles.

The first option suggests that mankind coexisted with dinosaurs before and after the Flood until dinosaurs were all extinct. There were no catastrophic events after the Flood, so since dinosaurs survived the Flood and were alive for God to use them in His illustration for Job, in that case, they are most likely still living amongst us, but their life span is too short because of a lack of oxygen and a stronger force of gravity for them to grow large again. Remember that dinosaurs had to adapt to the changes in the post-Flood environment as well as other creatures, including humans. Since humans survived the change, then so too did dinosaurs. Since dinosaurs were adapting to the post-Flood environment, they would gradually have a reduced longevity of life as humans did. Additionally, since all dinosaurs are reptiles, we know that reptiles continue to grow as long as they are alive. In pre-Flood conditions with high oxygen concentrations, weaker net gravity, and higher buoyancy, the dinosaurs could live 900+ years, constantly growing larger and taller. But in post-Flood conditions, with lower oxygen concentrations and higher net gravity and a weaker buoyancy force, the dinosaurs would have gradually adapted to the conditions as all life forms did. Their life expectancy would have decreased proportionately in the same way. Imagine a dinosaur that in prior generations lived 900+ years and had all those years to grow. Now imagine a dinosaur in post-Food conditions only living 200 years at the time of the book of Job, and then imagine a dinosaur only living 15–25 years today. We would call them lizards, tuataras, Komodo dragons, and crocodiles.

Review: The Bible describes dinosaurs about six to nine generations after the Flood in the book of Job. Dinosaurs could still be living amongst us, but they would just be much smaller than their ancestors. We wouldn't call them dinosaurs, but reptiles.

There seems to be only one logical conclusion for what caused earth's atmosphere to have 50% higher oxygen concentration in the past versus today. That is there was more usable land to grow more vegetation to produce more O2. And since the oceans are the primary limiting factor, then there were no oceans. Only the global flood could have caused a decrease in vegetation, increased oceans, massive polar ice caps, deserts, uniform layers of the crust, petroleum reserves, petrification, and coalification. That much water coming upon the earth from above and deep caverns below the crust would have covered the globe and formed vast oceans. Only the Flood could have reduced the high heat from the asteroid impacts, volcanoes, and violent tectonic plate movements and blocked the sun to produce the Ice Age. And with the loss of the protective layer of water above the atmosphere, the sun would then cause the deserts to form as they baked near the equators. Only the pre-Flood canopy could store up enough ocean waters to allow enough usable land and generate ambient global temperatures for vegetation to grow globally and produce higher concentrations of oxygen in the atmosphere, which caused life forms to grow larger and faster and to live longer.

Chapter Summary: Oxygen concentration in the atmosphere was 31% before the Flood. The Flood resulted in polar ice caps, deserts, and oceans that physically limited vegetation growth and subsequently reduced O2 concentrations to 21%. This caused life to suffer and shortened life spans. We see the longevity of genealogies reduced with the Biblical record, and thus, animals' longevity also suffered the same reduction in life span. This significantly reduced the quality and quantity of life. Atlantic ocean core samples reveal that the desertfication of the Sahara occurred around the time of the Flood.

<u>Group Discussion:</u>

1. Since the Bible associates life with the breath, how do you think that correlates with reduced oxygen after God's judgment via the Flood?

2. As the population grew after the Flood and consumed more oxygen, and considering that the oceans, deserts, and polar ice caps now limit oxygen production, then do you see the link with gradual reduced oxygen concentration in the atmosphere, with reduced longevity of life after the Flood?

3. Isaiah 56:20 describes humans living long lives again once the new heaven and new earth is created. Do you suspect God will use an increased oxygen concentration to fulfill this prophecy?

Chapter 6
Land Was More Plentiful in the Past

A premise of this book is that the earth's net gravity was approximately 13.5% to 21.5% weaker in some past millennium and oxygen concentrations in the atmosphere were 50% higher than today's current values. And those two changes caused life forms to be enormous in size and to live a long time. What did the earth look like when gravity was weaker and oxygen levels were higher?

For the above premise to occur, with 13.5% to 21.5% weaker gravity and 50% higher oxygen concentrations, there needs to be more land and less ocean water covering that land. The more land that is accessible for vegetation growth, the more oxygen is produced by vegetation. The ocean water also needs to be stored in a place that somehow reduces gravity and blocks the sun's harmful rays. The oceans physically limit the amount of vegetation that can grow and thereby directly reduce the oxygen concentration in the atmosphere.

For dinosaurs to live long and grow to enormous sizes on porous bones, gravity needs to be ~21.5% weaker than today's current value. Either the earth's mass was less, or Earth needed to have spun much faster, or buoyant force needed to be much larger, or some combination of that. For dinosaurs and humans to have lived such long lives, almost 1,000 years, oxygen concentrations needed to be around 50% higher than today's current value, which means more land needed to be exposed to grow more vegetation and produce more oxygen.

The solution that solves the above problem and fulfills both requirements of this book's premise of weaker gravity and higher oxygen is removing most of the earth's oceans and placing some of that water around the atmosphere and some in deep caverns under the crust (surface) of the earth, leaving maybe one-fourth remaining as seas. This would weaken the earth's gravity because of the buoyant force and increase the earth's oxygen concentration by increasing the exposed land for vegetation growth.

In addition, a solution that aids the premise in reducing gravity is a faster spinning earth. We already established that the earth spun faster near the time of creation, with Adam and Eve experiencing ~17 hours days.

Also, with some of the ocean water stored in the canopy surrounding our atmosphere, this would create too much atmospheric pressure for clouds to form and create a greenhouse effect and subsequent ambient temperatures on a global scale. This would convert deserts and polar ice caps into lush vegetation. Since large global changes in temperatures are required for the jet stream, there wouldn't be a jet stream and subsequently no hurricanes or tornadoes. The jet stream didn't start until after the Flood (Gen. 8:1). With the high atmospheric pressure from the canopy, evaporation would condense into a mist (Gen. 2:6), bypassing cloud formation.

Review: Removing some of the ocean water and placing 33–50% of that water surrounding our atmosphere and the other 25–33% in deep caverns under the surface of the earth would result in about ~21.5% weaker gravity and a 50% increase in oxygen concentrations in the atmosphere.

Is this fantasy? No, even scientists hypothesize that the water on Earth came from outer space. They suggest that several comets possibly impacted Earth to deliver the water or that asteroids rich in water impacted the earth and delivered the water. The Bible records a similar premise that water came from outside the earth. First, let's go back to the beginning knowing that the earth was formed out of water and by water (2 Peter 3:5).

On the first day of creation, all the matter in the universe was created. The matter was engulfed by ocean water—the deep waters (Gen. 1:2). Then during the second day of creation, God expands the universe and forms the atmospheres (Gen. 1:6–8) with "waters which were below the expanse from the waters which were above the expanse." Thus, our atmosphere, once had water above and below. Then the waters below the atmosphere (Gen. 1:9–10) "He called seas." Notice that God made seas, not oceans. What is the difference? An ocean has about 20 times more volume of water than seas. Job 28:14 illustrates that the seas and the ocean—deep, are different: "The deep says, 'It is not in me'; And the sea says, 'It is not in me.'" The deep is the ocean.

Let's now turn to Gen. 7:11: "All the fountains of the great <u>deep</u> burst open, and the floodgates of the sky were opened. The rain fell <u>upon the earth</u> for 40 days and 40 nights." Notice that the rain fell upon the earth, not on a particular city or region or continent. Also notice that it rained 40 days straight. All that water came from the canopy that was surrounding the atmosphere from when God created the atmosphere "between the waters" on the second day of creation. Furthermore, the deep burst open and poured its water onto the top soil to form the deep oceans mentioned in Job 28:14.

The global flood explains where the oceans came from, how the polar ice caps formed, when deserts were created, why mankind no longer lives 900+ years, and it also accounts for all the layers of the earth's crust, sea fossils at the tops of mountains, wind, and so much more. For this global flood to occur and form the oceans, a huge amount of water had to be stored above the atmosphere for that water to come out of the heavens and rain on earth for 40 days and 40 nights. Also, massive amounts of water had to be stored in deep caverns under the crust of the earth.

When the Bible uses the words of *all the fountains of the great deep burst open*, it is referring to violent and massive volumes of water erupting through the crust. Now, what happens when deep caverns inside the earth violently burst open? Earthquakes occur, volcanoes erupt, and tectonic plates shift, and a lot of dirt gets mixed into the water. Earthquakes are a result of volcanic activity and tectonic plate movement. Since volcanoes release a lot of moisture into the air, then the fountains of the great deep could be volcanoes. Volcanoes sure appear as fountains of the great deep that burst open today. And the source of the force to push massive volumes of water out of great deep caverns was the tectonic plates moving fast. Water being forced out as fountains and bursting from within the earth would be accompanied with massive amounts of dirt, sediment, silt, clay, rock, mud, and so on. The flood waters would have high turbidity (not clear water). And it takes time for soil in the water to settle —the Bible records that exact duration of the entire flood saga as almost one year.

It is important to note that the amount of mass divided by the volume of an item determines its density. The greater the density of sediment, the faster it sinks in a medium such as water. And the lesser the density of sediment, the slower it sinks in a medium such as water. Thus, the degree of mass compaction of the matter mixed in the flood waters determined the rate of descent from floating to settling on the ocean floor. This degree of mass compaction is called density. Therefore, density determines the rate of descent while matter is floating in a medium such as water. All things will settle according to their density, based on the laws of physics. And all that dirt, rock, silt, soil, sediment, clay, and the like mixed in the turbulent and violent Flood settled according to density and formed layers, which is exactly what we have today. Aiding the segregated settling of the different soil—based on their densities—was a twice daily global tide (this process is called liquefaction).

Since the global flood was a judgment from God on mankind and earth and since this was not from cloud formations, it's safe to say that the rate of the rainfall would have been much higher than the severest thunderstorm. If all the mountains and valleys were smoothed out on the spherical earth, then the water on earth would be ±1.5 miles deep, which is ~7,920 feet. Then if we estimate that 20% of the oceans existed as the seas, that accounts for 1,584 feet of the water, and 50% of the oceans came from the canopy and ice comets, that accounts for 3,960 feet of the water, and 30% of the oceans came from the fountains of the great deep, that accounts for 2,376 feet of the water. This accounts for where all the water on the earth may have come from. Perhaps it could have been 50 inches of rainfall per hour for 40 days and nights. This equals 4,000 feet of water (not including ice comets); this is not enough water to cover the tops of mountains today, but remember that the mountains and valleys rose and sank after the Flood (Psalm 104:5–9). Thus, we do not need water to cover the existing height of mountains, but the lower height of mountains back in the Gen. 7 time period. All the great mountains arose after the Flood, when the valleys sank and the mountains rose to peaks the size of Everest, and that is why all mountains have seashells on top of them. Psalm 104:6–8: "You covered it [earth] with the deep as with a garment, the waters were standing above the mountains. At your rebuke they fled, at the sound of Your thunder they hurried away. The mountains rose, the valleys sank down to the place which You established for them." The likely impact of a frozen comet impact near the Arctic Circle, which would have been able to freeze woolly mammoths in an upright position with undigested food in their stomach (several such specimens have been discovered in Siberia).

We know this was a global flood and not a local regional flood because archeologists have found seashells at the tops of the highest mountains. And God's Word (Gen. 7:19–23) directly says that all the high mountains everywhere under heaven were covered by the waters, and all flesh that moves had died. And in Gen. 8:3: "The water receded steadily from the earth, [not regionally] and at the end of one hundred and fifty days the water decreased. [A local flood doesn't take 150 days to decrease, but this global flood took almost a year to recede.] In the seventh month . . . the ark rested upon the mountains of Ararat." (A local flood would not put a ship the size of a football field on top of the mountains of Ararat in Turkey. Those mountains are high, and Noah's Ark has been found in those mountains). *The water decreased steadily until the tenth month, the* **tops of the mountains became** *visible*. The flood started on the second month, and in the 10th month, the mountain tops were visible! The picture shows the fossil remains of a wooden vessel with a collapsed roof and with exact proportions as recorded in the Bible for Noah's Ark. *Photo credit: Ron Wyatt: Wyatt Archaeological Research, Mary Nell Wyatt, and Richard Rives.*

Gen. 8:6–9, makes is quite clear that this was a global flood with, "the water was on the surface of **all the earth**." At this point, if you still refuse to believe this was a global flood, then you must question your faith in God because God declared it in His testimony. You can take it at face value and look like a fool before mankind, or you can reject it to pacify nonbelievers, but don't think the Bible doesn't clearly record a global flood.

For this much rain to come down and fountains of water to burst open, this indeed would have formed the oceans as we know it. This means that before the Flood came, the earth had a lot more land available for vegetation growth that produced massive amounts of oxygen, 50% more than today. If you don't think a canopy of water surrounding our atmosphere is possible, consider Saturn has ~26 million times more water surrounding it as rings than on the entire earth. Considering Ecclesiastes 1:9 "That which has been is that which will be," and Rev. 21:1 "A new earth . . . there is no longer any sea." The concept of more usable land before the Flood is also supported by future prophesy.

Chapter review: The Flood brought so much water upon the earth that it created the world's oceans. There were no oceans before the global flood. Before the Great Flood, only seas existed. Prior to the Flood, there would have been extra land that would have had extra vegetation, and that extra vegetation would have produced extra oxygen—50% more than after the Flood.

<u>Group Discussion:</u>

1. How does the concept of the earth having 70% land and 30% water before the Flood change your perspective?

Chapter 7
Meteors, Asteroids, and Comets

In the burgeoning first day of the universe, the space between forming planets, stars, solar systems, and galaxies was filled with matter in the form of hot water and soil. After the rapid expansion of the universe, galaxies, and solar systems on the second day of creation, then the space between the galaxies and solar systems was filled with colder asteroids, meteors, and comets, much more than we observe today. Most of those floating objects were drawn into the nearest planet's, moon's or star's gravitational force. On planets that are active, such as Earth, the impact craters have been covered up from erosion, tectonic plate shifts, forest, oceans, and so forth. But those planets and moons that are no longer active, such as our moon, show a vast array of impact craters, indicating that there was heavy activity in past millennia by impacting asteroids after its formation. Most of the asteroids and meteors that were circling the sun between the planets have been drawn in by those planet's gravity and have been absorbed by now. But there are still two belts of asteroids and meteor clusters within our solar system, called the Kuiper Belt and Main Asteroid Belt.

The Kuiper Belt is located between Neptune and Pluto and consists of millions of asteroids that encircles our solar system and may be the last remnants of how our solar system existed with free floating molten masses before they coalesced into planets. And the remaining cooled asteroids are what is left over of what floated all throughout our solar system, not just encircling outside of it. *Image credit: NASA and – I believe – G. Bacon (STScI).*

The Main Asteroid Belt lies between Mars and Jupiter and is a cluster of asteroids and meteors. This belt could represent the frozen leftover material that didn't coalesce into a hot celestial body during the first day of creation. And when the hot universe expanded on the second day of creation, the molten matter that was caught in no man's land, cooled into asteroids and comets. *Image credit: NASA/Goddard Space Flight Center.*

Why have a chapter about asteroids, meteors, and comets? Because they played a significant part in the formation of the celestial bodies of our solar system, may have initiated the global flood of Gen. 7, and will be used by God to fulfill the end time prophecies. Both creationists and evolutionary cosmologists accept that in the beginning, hot molten matter coalesced into the nearest celestial body to form planets, stars, or moons and that the two asteroid belts are the remnants of that period. They just differ on when this process occurred and how long this process took.

And both creationists and evolutionary cosmologists accept that after life began on Earth, there was another period of asteroid, meteors, and comets bombarding the earth. The difference is that the evolutionists' view suggests a major event occurred 65 million years ago, and in the creationist view, the event occurred at the initiation of the Gen. 7 catastrophic flood. When I read an evolutionist explain how some massive asteroids hit the earth 65 million years ago that caused a mass extinction, I say, well the same scenario played out at the beginning of the global flood. Evolutionists contend that the impact would have sent debris 400 miles wide and a shock wave 1,000 miles wide. A creationist agrees with the synopsis. Evolutionists further say that the terrain burned and left a layer of carbon-based ash and soot. Agreed. The heat from the blast would have increased global temperatures to more than 100°F plus and would have killed off what the blast zone didn't. Also, there would have been acid rain in the blast zone. All these suppositions are in agreement with the creationist view of what was going on in the initial stages of the global flood of Gen. 7. Evolutionists further say that this asteroid impact event plunged the globe into an ice age. That's agreed to as well. If there is so much agreement here, then why do we differ on when it occurred? Because of the foundational premise for interpreting the observable evidence. If the Bible is wrong with the genealogical time lines, then there is no God. If the Bible is correct, and God created everything in six literal days, and the conditions changed with a literal global flood, then evolution is wrong.

Each time that a comet orbits about the sun, it loses matter. Water and debris are ejected off the comet until there is no longer any ice left on the comet. This is demonstrated by its tail that points away from the sun. And since there is no observable evidence of comets being formed, or existing comets having frozen water added to them, then we may conclude that each time we observe a comet with its tail, then it is strong evidence of a young solar system and galaxy. If the universe was millions of years old, the comets would have lost all their water from orbiting about the sun. With this hypothesis, there should be a reduction in the size of the tail representing ice being ejected off a comet with each orbit around the sun. And indeed that is what cosmologists have observed. Evolutionary cosmologists counter this theory with the Oort cloud; just outside the Kuiper belt, the Oort cloud of icy material adds ice to each comet. However, this has never been observed; thus, it leaves the reality of science and enters the realm of faith-based religion. *Image credit: Wikipedia.org/comets.*

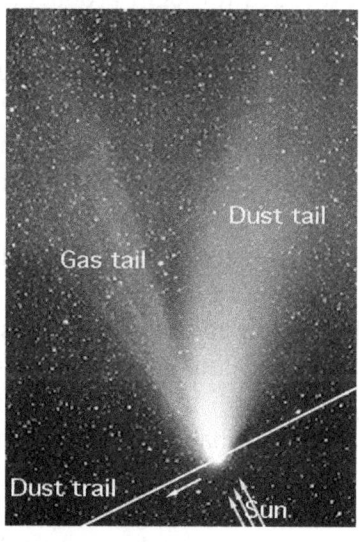

Let's work through some concepts about asteroid impacts and determine how their impacts have shaped life on earth. Each of the comets and asteroids that impacted Earth would have adversely affected the planet's spin velocity and added to Earth's mass and increased gravity. For the most part, the effect of adding mass—after creation—to increase gravity was negligible, but there were large impacts that did have a noticeable change on life on earth. A hypothesis of this book is that asteroids impacting the earth initiated the Flood by colliding into the frozen arch support structure.

As the asteroids passed through the canopy and impacted the earth, this could have slightly disrupted the spin of the earth and reduced the centripetal force, thereby increasing the net gravitational effect.

When the earth was formed, there were many more asteroids, meteors, and comets. How do we know this? Well, every time a comet, asteroid, or meteors collides into a planet, moon, or sun, it is gone forever. And we don't see any factories producing them. Our moon, though it is small compared to Earth and has a weaker gravitational pull to attract asteroids, has thousand of impact craters; thus, it can give us clues about what has happened to the earth. The more mass a planet/moon/star has, the greater the gravitational pull, and the greater the attraction of asteroids, comets, and meteors to it. We know that our moon has been hit thousands of times, and given its smaller size and weaker gravitational pull, we can presume that it has been hit far fewer times than the earth. The moon is no longer active, so its impact craters are not covered up by erosion, earthquakes, wind, oceans, and so on. Therefore, the earth has been hit more than the moon, but the impacts are not as obvious. In addition, Jupiter is 1,321 times larger than Earth. Thus, given Jupiter's larger size and gravitational force, it would absorb many more asteroids than Earth. Likewise, Neptune is 58 times larger than Earth, and Uranus is 63 times larger, and Saturn is 763 times larger than Earth, and the sun is 1.3 million times larger than Earth. Put all this together and it means that there were ~hundreds of millions more asteroids, comets, and meteors many millennia ago as compared to today.

The point of talking about asteroids, comets, and meteors impacting the earth is that something had to trigger or initiate the global flood. The frozen external portion of the canopy would have provided the necessary support to maintain its set altitude, and additional assistance to maintain the canopy up above the atmosphere would come from centripetal force. I hypothesize that asteroids crashed into the canopy, breaking the frozen external support arch, then impacted the earth, slowing the earth's spin velocity, and initiated the Flood. This crossed a threshold and allowed gravity to bring the canopy down upon the earth. There came a point when the spin creating centrifugal force and the exterior frozen arch of the canopy was overmatched by the earth's gravity. A threshold is a point exceeded to begin producing a given effect or result. For example, when water is 33°F, it is just cold and not freezing. When water crosses the threshold to 32°F, water freezes. If gravity hadn't pulled the canopy down to earth, the canopy would have eventually formed rings just like Jupiter, Saturn, Uranus, and Neptune. But gravity overpowered the centripetal force and pulled down the canopy. The canopy came into the atmosphere, melted, and rained for 40 days and 40 nights. What caused the earth to lose enough spin to cross the threshold? And what caused the frozen external archway of the canopy of water to give into gravity?

A series of large asteroids crashing into the frozen external arch of the canopy of salt water and hitting the earth would have caused the earth to slow and caused a net increase in gravity. In addition, the moon continuously takes angular momentum away from the spin of the earth, thereby slowing the earth and thus increasing gravity. This wouldn't have to be a large reduction in the spin of the earth—just enough to cross a threshold. There are craters on the earth that are large enough to indicate an impact capable of altering the rotational velocity of the earth. The so-called "dinosaur killing asteroid(s)," such as the one that hit the earth at the Yucatan Peninsula, could literally have been a dinosaur killer, a life killer, and initiated the Flood.

Is there physical evidence supporting the theory of an asteroid impact large enough to corroborate my theory? Yes, there is ample evidence of large enough impacts to be the legendary "dinosaur killer." One of the most telling pieces of evidence is a thin layer of iridium. *Photo credit: sciencebuzz.org.*

What does the picture show? For one, when trees get blown over by asteroid impact zones and massive amounts of dirt cover up the trees, then vegetation turns to coal. So the impacting evidence above the coal line is in harmony with the Bible. There is also a layer of iridium concentration. Where does iridium come from? It comes from asteroids. Asteroids have iridium, but the earth doesn't have iridium unless it is brought from an outside source. A layer of iridium like this in several locations around the globe indicates that at one time, many large asteroids impacted the earth around the same time frame. How do we get a second layer of coal above the asteroid impact layer? Tree bark gets separated from the trees during impact trauma, and trees rubbing together knock the bark off, which sinks after being dislodged from the trees. The bark combines with other vegetation and trees and eventually forms the bottom layer of coal. The higher second layer of coal could have come from trees and vegetation that floated for a couple of weeks and sank on top of the impact layers, thus causing the second layer of coal. Then the muddy waters from the Flood would have settled on top of the vegetation over the next year. The different sediment would have settled by density according to the laws of physics, hence creating the mudstone layer on top. This could explain the reason for these similar layers around the world.

Coal is created from botanical life covered with pressure, heat, and moisture. A global flood is a perfect scenario to convert botanical life to coal. It has the pressure, heat, and moisture necessary to accomplish coalification. At the beginning of the 40 days of rain, there was a massive amount of heat generated on the planet from several major events, such as asteroid impacts, which fractured Pangaea like an egg and released 1,650 years of stored-up tectonic plate energy, causing the tectonic plates of Pangaea to break apart at fast speeds from 0.06 to 0.5 mph. The Bible mentions a book called Jasher in Joshua 10:13 and 2 Samuel 1:18, but since the only copies of Jasher are in a fragmented incomplete form, the early patriarchs elected not to include the book in the Bible. But potentially, the book of Jasher may give a clue as to what caused the earth's only land mass, Pangaea, to crack into the different continents we see today. Jasher 6:11: "And on that day the Lord caused the whole earth to shake, and the sun darkened, and the foundations of the world raged, and the whole earth was moved violently, and the lightning flashed, and the thunder roared, and all the foundations in the earth were broken up." Food for thought: this indicates the violence associated with the Genesis flood as Pangaea was broken apart during that saga.

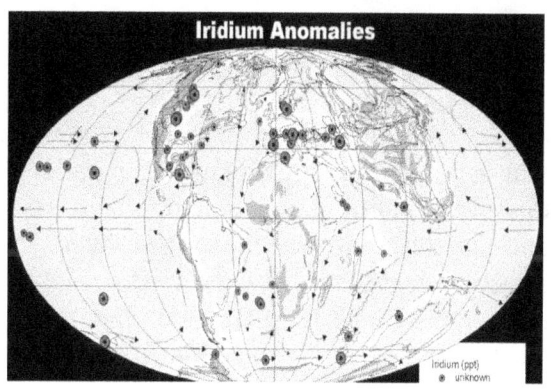

The force from the shifting of tectonic plates caused water from deep caverns to burst out from under the crust of the earth, sending tons of sediment into the air and eventually into the floodwater. While the massive tectonic plates were moving, they caused fountains of lava to erupt from hundreds of volcanoes around the globe from multiple eruptions during the Flood event. Remember that the floodwater of Gen. 7 would have been filled with massive amounts of dirt/soil because the deep caverns of the deep under the crust of the earth had burst open. And what caused the deep caverns of the earth to burst open? Asteroid impacts and tectonic plate movements. And when the sediment in the water finally settled, the soil settled in layers. And that is exactly what we find on top of the iridium layer, layers of sediment. *Image credit: www.sites.google.com/site/thatdinothing.* The picture shows the locations of iridium. Iridium deposits are evidence of asteroid impacts. The Flood saga may have needed a comet of frozen ice to impact a polar region to cool down the heat generated from the catastrophic event. There are mammoths frozen in the upright position, with undigested food in their digestive tracts. This indicates that a frozen comet of water could have impacted the Arctic Circle to explain such rapid cooling. A comet would have cooled the intense heat of the Flood.

Do evolutionary geologists support my hypothesis? Yes and No. No: they don't believe these impacts happened 4,500 years ago during the Genesis Flood. Yes: They believe there were asteroid impacts but that they occurred hundreds of millions of years ago when they wiped out the dinosaurs.

Does the Bible support asteroids hitting the earth? Yes, it supports an asteroid event in Joshua 10:10–11. Let me set the scene. Israel is in a battle against wicked people that have violated God's Word. The general leading Israel is Joshua, and God tells Joshua, "Do not fear them, for I have given them into your hands; not one of them shall stand before you." Then listen to what God does:

> And the LORD confounded them before Israel, and He struck/slew them with a great slaughter at Gibeon, and pursued them by the way of the ascent of Beth-horon and struck them as far as Azekah and Makkedah. As they fled from before Israel, they were at the descent of Beth-horon, the LORD threw large stones from heaven on them as far as Azekah, and they died; more who died with [stones] than those whom the sons of Israel killed with the sword.

Could an asteroid hit the earth and increase the length of the day? Yes. If a large asteroid hit the earth in the opposite direction of the earth's spin, the earth could slow its rotation and subsequently lengthen the day—similar to what occurred to Venus. Cosmologists theorize that Venus was hit by a series of asteroids that caused a reversal of its spin. Now, when the impact on Earth landed on a tectonic plate, it could have dislodged that plate and have the plate slide in the direction of the vector of force. And if that direction was toward the west, then that plate would slide on the surface of Earth parallel with the sun's trajectory, causing the occupants standing on the surface to view the sun as though it was standing still. But since all the oceans, seas, and all dirt on the planet were still spinning at the same speed as the earth, the energy (called inertia) of all that matter would force that momentarily moving tectonic plate to continue rotating with the earth. Has this ever happened to earth? There is a record in the Bible where a day was prolonged. Joshua 10:12–14:

> Joshua spoke to the LORD,'O sun, stand still at Gideon, and O moon in the valley of Aijalon.' So the sun stood still, and the moon stopped, until the nation avenged themselves of their enemies . . . And the sun stopped in the middle of the sky and did not hasten to go down for about a whole day. There was no day like that before it or after it, when the LORD listened to the voice of a man; for the LORD fought for Israel.

This is from the perspective of the viewer on earth, so the sun didn't stand still, but the tectonic plate that they were standing on moved parallel with the sun's normal trajectory of setting in the west. We do this every morning and evening when we say, "Oh what a lovely sunset/sunrise." The sun didn't rise, and the sun didn't set; the earth rotated. We don't say, "Oh what a lovely Earth rotation." That is the same concept here. How did God cause the sun and moon to seemingly stand still? Perhaps it was by momentarily causing a tectonic plate to slide along the path of the sun from the earth spinning. How did God move the tectonic plate, which seemed to cause the sun and moon to stand still? It was from asteroid impacts. Just prior to Joshua recording that the sun stood still in Joshua 10:12–14, he describes God striking and killing the wicked people in verses 10–11 with:

> The LORD confounded them before Israel, and He slew them with a great slaughter at Gibeon and pursued them . . As they fled . . . <u>the LORD threw large stones from heaven on them</u> as far as Azekah, and they died; <u>more died from the stones</u> than those whom the sons of Israel killed with the sword.

This is exactly what we would expect to read if an asteroid caused a tectonic plate to slide along the path the sun moves. It indicate that the "<u>large stones from heaven</u>" were asteroids and meteors, which were from heaven (space). Those asteroids killed more people than Joshua and his armies did by sword. This fits perfectly with asteroids and meteors hitting the earth, killing a bunch of people. And the force from the impacts caused a section of the surface crust of earth to slide and mirror the rotational spin of the sun to seemingly prolong the day, potentially by sliding the entire tectonic plate on top of magma until the inertia of all the matter on earth and the spin of the iron core and inner mantle caused the earth's crust or that particular set of tectonic plates to resume its rotational movement with other crust. There is scientific evidence to support a literal reading of the text that says the day stood still from the viewer's perspective on earth.

The Bible also supports asteroid events in the near future in Revelation 6:13. "The stars of the sky fell to the earth, as a fig tree casts its unripe figs when shaken by a great wind." If a large enough asteroid hit the earth, this would slow the spin and cause the earth to wobble off its 23° axis, much like when a gyroscope slows and starts to wobble. Does the Bible support the notion that the earth will be hit by such a large asteroid that the earth will wobble? Isaiah 24:18b: "For the windows above are opened, and the foundations of the earth shake. The earth is broken asunder, the earth is split through, the earth is shaken violently. The earth reels to and fro like a drunkard and it totters like a shack." When does this happen? Revelation 16:17–20:

> Then the seventh angel poured out his bowl upon the air, and a loud voice came out of the temple from the throne, saying, 'It is done.' And there were flashes of lightning and sounds and peals of thunder; and there was a great earthquake, such as there had not been since man came to be upon the earth, so great an earthquake was it, and so mighty, the great city was split into three parts, and the cities of the nations fell . . . And every island fled away, and the mountains were not found. And huge hailstones about one hundred pounds each, came down from heaven upon men.

The Scriptures indicate that a large asteroid hits the earth so hard that it splits the earth through and causes the largest earthquake ever. Note that *asunder* indicates the splitting of one whole into two parts. This is used, for example, in the vow of marriage, "What God has joined together, let no man split asunder." A likely location is at the Mid-Atlantic Ridge that stretches from the North Pole region to the South Pole region and is a divergent zone. At this Mid-Atlantic Ridge, the crust is pushing up and away from itself, which causes the African and European tectonic plates to push into Asia and the North American and South American tectonic plates to push into the Pacific plate. It should be noted that when the earth splits asunder, all the oceans will pour into the hot mantle and core, causing the water to shoot up and surround the atmosphere, reconstituting the canopy again. And while the water is shooting up into space, this will create a jet engine of force on a global scale that will put the earth back into a fast spinning orbit just like before the Flood. This explains how in Rev. 22:1, there is no sea.

Review: Asteroids have hit the earth, and the Bible pronounces that they will hit the earth again. Both God and geologists agree that asteroids have impacted the earth, but they disagree as to when those asteroids impacted the earth. It is possible for asteroids to hit the earth and reduce the spin of the earth or cause a tectonic plate to slide. It is probable that the "dinosaur killing asteroid(s)," such as the one that hit the earth at the Yucatan Peninsula, could have initiated the Flood.

Group Discussion:

1. Since the Bible records that asteroids have hit the earth and will again hit the earth in the future, all in divine judgment, do you think that asteroids initiated the judgment of the Flood?

2. Does the explanation of the famed "prolonging of the day" during Joshua's time increase your faith in the Bible and why?

Chapter 8
Earth's Spin at Origins

In the beginning, on the first day of creation, when God started rotating the formless mass of water and dirt that was to be Earth, that rotational speed was faster than the current rotational speed of the earth. This means that the evening and morning of the first day occurred slightly faster and in shorter time than the current 24 hours it takes for one rotation of the earth.

How do we know this? The moon is one reason we know this. The moon's gravity takes angular momentum away from the earth's rotational velocity, and the moon uses the captured energy to move away from the earth at 3 inches per year. Each ocean tide, twice a day, is caused by the moon's gravity. Each time the ocean tides occur, angular momentum is taken out of the earth's rotational velocity. This is to say that the moon is slowing the earth down by means of gravity. When the moon was closer to the earth, the moon's gravitational pull would have had a greater effect on the earth's ocean tides. With greater oceans tides, more angular momentum was taken out of the earth's spin. The more angular momentum taken out of earth's spin, the greater the reduction in the earth's spin.

Therefore, when the moon was closer to earth, the moon slowed the spin of the earth to a greater degree. Putting all the puzzle pieces together, we discern that the earth spun faster in past millennia. With a faster spinning earth, the length of one day was shorter. This begs the question: "How long was a day when dinosaurs roamed the earth?"

Currently, the earth is slowing 2 milliseconds per 100 years. But this rate of reduction of speed is not constant. In prior centuries, the rate of reduction would be larger. Why? Because the moon would have been closer and its gravitational pull would have caused greater ocean tides, which would have taken more speed out of the spinning earth. The further back in time we go, the greater the reduction of spin exponentially. In addition, large asteroid impacts would have reduced the earth's spin velocity. Also, when large asteroids hit the earth to initiate the Flood and the breakup of Pangaea, the large tectonic plate movements would have also taken away angular momentum from earth's spin. Putting these pieces of the puzzle together, an estimate of the length of day at the time of Adam and Eve, some 6,000 years ago, would be approximately 15–17 hours long.

The velocity of the earth's spin when Adam and Eve and the dinosaurs roamed the earth would have been around 1,606 mph, which is ~60% greater than the current rotational velocity of 1,037 mph. This would also have reduced the gravitational effect from centripetal force. One cannot go too far back in time with aging the earth, otherwise, ocean tides would be too great and the earth's spin would be too fast. This supports a young Earth scenario.

An interesting note regarding the spin of spiral galaxies is that the calculated speed of their rotations produces a greater escape velocity than the gravitational attraction that holds it together. This means that the age of each spiral galaxy represents a young universe, because with an old universe, each spiral galaxy would have lost their distal appendages of solar systems and the concentrated core of each spiral galaxy would have dispersed by now. Evolutionary cosmologists counter this theory with the argument that dark energy holds the spiral galaxies together. We really don't know much if anything at all about dark matter. All hypotheses regarding dark matter are educated guesses.

Review: The earth spun faster in past millennia and subsequently had shorter days and weaker gravity as a result of the faster spinning earth because of centripetal force.

Group Discussion:

1. With the knowledge that the duration of a day was shorter the further back in time one goes, how does this affect your interpretation of the day in the Genesis creation account?

Chapter 9
The Flood

This event is so instrumental in changing the conditions on the planet and life on Earth that it needs some attention. Discerning what the Flood caused and changed will reveal what the conditions and life on Earth looked like from creation till the Flood. One can use the Bible's record to interpret observable evidence to understand from coal mines, shallower salt mines (the deeper salt mines were formed during creation), oil reserves, petrified wood, dinosaur fossils, layers of soil in the crust, deserts, polar ice caps, the continental shelf, seashells on top of mountains, ancient cultures with flood stories, large asteroid impact craters, tectonic drifts, mountains, and valleys. All that we see today can be explained in the context of the Flood, either directly or indirectly, including the notion that mankind once lived 900+ years while dinosaurs roamed the earth.

We'll go through a description of the Flood based on the Bible, including hypotheses based on deduction from observable evidence and science and then attempt to break down the major points throughout this book, with a goal of using the Flood to shed more light on what the environment and life were like on the planet before the Flood at the time of the origins of life.

~6,000 years ago and ~1,650 years pre-Flood:

The Bible records that God stored water in three locations at the time of creation to allow optimal living conditions on Earth. Two of those locations were later used for judgment with the Flood. One location of water storage was in the deep caverns just below the surface (Gen. 7:11). The second location of water storage was in the canopy hovering above the atmosphere (Gen. 1:6–8). The third location of water was the seas (including freshwater lakes). The potential percentages of water in these three locations that currently make up our oceans and seas, range from 33%–50% in the canopy, 25% to 40% in the deep caverns, and 20% to 33% as seas. This utopian setup that God created was called Eden, from the time of creation until the Flood. There were no large oceans, but there were small seas, and the exposed land was covered in lush vegetation. There were no polar ice caps or deserts, but plenty of vegetation. There were no four seasons, just spring. Resulting from evaporation and high atmosphere pressure, the hydrological cycle skipped clouds and rain and went straight to a mist that rose from the ground and watered the whole earth (Gen. 2:5–6). With no polar ice caps and no hot equator, there was no jet stream (Gen. 8:1). The moon was closer and larger, which caused sea tides to be greater. The earth spun faster, which shortened the day to around 17 hours. As a result of this canopy of water (and other factors) hovering above the atmosphere, a large increase in the buoyancy effect reduced net gravity by around ~20%–25%. There was a supercontinent called Pangaea (Gen. 1:9 and 10:25) that potentially covered 70%–80% of the earth's surface, which was covered by vegetation, which produced large quantities of oxygen, causing life to thrive. The result of weaker gravity and greater oxygen allowed dinosaurs to grow to massive heights and sizes (Job 40 and 41) despite having porous bones and allowed humans to live to 900+ years of age (Gen. 5).

But then mankind sinned and set in motion an environment in which all of creation would groan and suffer (Rom 8:22) until the Messiah returns as a conquering lion to restore all of creation back to its pre-fallen state at origins (Isa. 11:6–10 and Rev. 22). At the moment of the first sin in the Garden of Eden, the movements of judgment began with the inner earth churning to set up future earthquakes and future volcanic activity and future tectonic plate movements. Though connected as Pangaea until the Flood, the potential energy began building up for the judgment at the global flood. Asteroids were colliding in distant space to set off chains of events for future collisions with Earth, yet the result of these movements and activities wouldn't be seen until the earth passed through the path of these asteroids and meteors. The canopy was morphing from a spherical shape to a disc-like shape from its spin with the earth and the pull of the moon's gravity so that occupants on the surface of the planet could see a visible arch in the sky.

Adam and Eve were obedient to God's command on the sixth day to multiply, and they probably had a hundred plus offspring by the birth of Seth (130 years after creation). There are estimates that the population from Adam to Noah, with people living 900+ years and having many more children, that the population was 7 billion people at the time of the Flood.

120 years pre-Flood:

Mankind became so wicked that the Bible records that some demons left their natural abode and bore a half-demon, half-human creature with the daughters of men and produced offspring called the Nephilim, who were mighty giants and men of renown. Gen. 6:1-8 says "that every intent of the thought of his heart was only evil continually. The LORD was sorry He had made man on the earth, and He was grieved in His heart. . . But Noah found favor in the eyes of the LORD." When God saw the wickedness of mankind, He said you have 120 years until judgment (Gen. 6:3). Noah and his three sons got busy building an ark that was 450 feet long, 75 feet wide, and 45 feet tall, with lower, second, and third decks (Gen. 6:15–16). This provided 15 feet for each deck, made out of gopher wood and covered inside and out with pitch (a petroleum resin) (Gen. 6:14). The total volume was 1.4 million cubic feet. Imagine the faithfulness of Noah and his family to build an ark for 120 years, when it had never rained one time.

Seven days pre-Flood:

Two of every kind of land and air creature entered the ark, one male, one female. This includes dinosaurs and other very large creatures, but excluded aquatic life. Utilizing young animals, especially when it came to reptiles, would have solved the logistical problems and would solve predatory problems. With seven days before the commencement of the Flood, Noah entered the ark, sealed by God (Gen. 7:4 and 7:11).

The first day of the Flood (±4,400 years ago around 2,400 BC):

As asteroids neared the earth, some collided with the moon and many collided with the canopy on their way to Earth. As a result of the multiple collisions with the canopy, the asteroids broke the frozen support arch structures of the canopy and caused it to slowly come down upon earth, as gravity gradually pulled it down. With raindrops having a terminal velocity of ~10 mph at Earth's surface and with the planet having a large equatorial bulge from its spin, the water would have come down over a duration of 40 days and nights, and not all at the same time. This would have been a slow descent that spanned 40 days from start to finish because the canopy had zero descent speed and had centripetal force from spinning ~1,600 mph. The asteroids, however, were traveling much faster, potentially around 30,000 mph, and they reached the earth's surface in seconds after contacting the canopy of water.

With Noah and family and the creatures safe in the ark, they would have heard the asteroids impact the earth way off in the distance. Surrounding those asteroid impact zones, life was devastated with collateral damage, debris, and fire. The shock waves of the multiple large asteroid impacts and thousands of meteors knocked down trees and caused fires and sent debris into the atmosphere. The initial layer of debris covered much of the vegetation and biomass, which would eventually form large coal mines and oil reserves where there was a thermal heat source.

The asteroid and meteor impacts were so severe that they fractured the crust of the earth as the asteroids impacted Pangaea. With 1,600 years of tectonic plate tension built up, the tectonic plates that made up Pangaea split in many regions, allowing the release of all that stored potential energy to be released as kinetic energy. Pangaea began to break apart, and the tectonic plates accelerated apart by hydroplaning on the deep caverns at high rates of speed. The distance between each tectonic plate varies, but we will use the South American and African plates to illustrate a point. They are traveling away from a divergent zone called the Mid-Atlantic Ridge—they are both roughly 1,750 miles away. Dividing that by 40 days of flooding plus the 150 days that the flood waters covered the planet and dividing that by 20 hours in an average day = ~0.5–1 mph average velocity that the tectonic plates were traveling during the Flood. To put this in perspective, humans walk at 2.5 mph. This separation of the tectonic plates forced the fountains of the deep to burst open.

The first week:

As Pangaea broke apart, this caused large volcanic activity. The Yellowstone's super volcano erupted a couple of times, and the Hawaiian volcanoes erupted around 50 times as the Pacific plate moved over its hot spot. Multiple large volcanoes erupting around the globe sent global temperatures rising. At the same time that the volcanic activity was occurring from the break up of Pangaea, the energy from the massive tectonic plates moving forced vast amounts of water stored in deep water caverns under the crust of earth to burst open (Gen. 7:11), spewing tons of superheated water and soil into the atmosphere. The deep caverns of water that the plates rested upon were the source of the reduction in friction that allowed the tectonic plates to freely slide apart and, in a sense, hydroplane. With the mixture of water bursting out of the ground, volcanoes erupting massive amounts of debris in the air, pockets of acid rain, and asteroids colliding on the earth, it was one chaotic planet—not suitable for life outside a protective ark.

The African tectonic plate separated from the South American tectonic plate, and the North American plate separated from the European tectonic plate from a central divergent zone called the Mid-Atlantic Ridge. While the planet was on the verge of too much heat that would kill all life, the canopy of water started raining upon the entire surface of earth. It quelled the fires, blocked the sunlight, and cooled the environment. The rain (without clouds) saved eight humans and two of every kind of animal on an ark larger than the size of a football field from dying from too much heat.

Second week:

With this much chaos, the waters would be very turbulent, and a boat that wasn't stabilized would capsize. This potentially explains the discovery of several 9-foot-tall, 2-ton anchor stones with attachment holes for ropes near the region of the ark. These anchor stones would have reduced the boat's lateral pitch and yaw and sway with the waves. With so many volcanoes erupting sulfuric particulates in the air, they caused sulfuric rain, which caused some ash, bone, teeth, coral, and shell fragments to break down to calcium carbonate and form the first layer of limestone.

As time passed, the tectonic plates continuously moved rather quickly compared to today's standards (at roughly 1 mph), moving land off old volcanic hot spots, off the pluming vent of magma, and moving new land over the magma vent. New eruptions occurred, and new islands formed, eventually forming a chain of islands. The old Yellowstone eruption remnants drifted west by southwest, and new land moved over the Yellowstone vent, and another eruption occurred with new debris, sending massive amounts of sediment and water into the atmosphere. All the while, the water was still bursting out of the many deep caverns just like Yellowstone, and rain was still falling from the canopy to cool the planet. The movements of the tectonic plates created large vortexes of swirling debris of sediment, vegetation, and biomass.

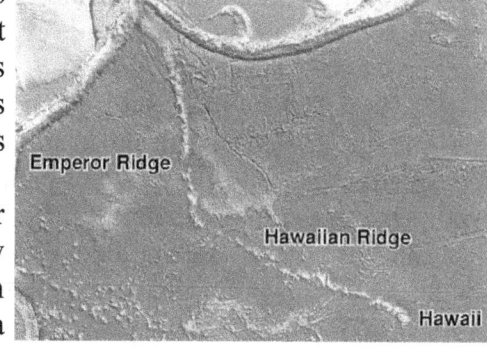

Within the Pacific plate, there were multiple smaller volcanic eruptions daily, as the plate moved north by northwest, creating a chain of tiny islands along the way from Hawaii toward the northwest. From all the magma displacement and asteroid impacts, the earth would have been a ball of flames, yet the sun's light was blocked by the rain and debris in the atmosphere, and the rains would have continuously quelled the flames and heat. The planet didn't overheat, but things buried near the heat either started the accelerated decay process of forming coal mines or oil reserves. By this time, the floor of the Pacific plate had seen nearly a 100 volcanic eruptions over the entire plate, with water levels rising and new islands forming by the day. The Pacific tectonic plate continuously moved north by northwest, and new islands formed each day as the subducted Pacific plate dove under the Asian plate. Noah's Ark was built in the Mesopotamian fertile crescent valley, and since there were no sails, no ores to row, no motor for mobility, and since there were potentially several 2-ton anchor weights that stabilized the boat's movement, the ark remained relatively stationary.

<u>One month:</u>

The rains still persisted, volcanoes continued to erupt, and superheated water continued spewing out of deep caverns under the crust of the earth, allowing for freer movements of the tectonic plates, which were still moving. Somewhere around this time, the Pacific plate took a turn and started heading west by northwest. This change of direction caused a temporary pause of volcanic eruptions, but only for a couple of days. The chain of newly formed islands continued to grow as the Pacific plate continued to move fresh land over vents of pluming magma. This activity formed the Hawaiian Islands in the Pacific and remnant chains of islands. A third Yellowstone eruption occurred through new fresh land that moved over the magma vent, blocking the vent only temporarily for this large eruption of superheated water, soil, ash, and flowing magma. The Indian tectonic plate slammed into the southern border of Asia and began the process of forming the Himalaya mountain range, though the mountains were originally below the surface of the waters by hundreds of feet.

Although the volcanic activity was still ongoing, the global rains and sun blockage prevented this catastrophic period from plunging the earth into a superheated inferno. And the temperatures came off their highs for the first time and cooled to an average global temperature of ~90°F on the surface, but below the water, the soil was hot, as liquid magma continued melting rocks. However, at the poles, snow may have been falling at an accelerated rate, with woolly mammoths freezing in an upright position with undigested tropical food still in their stomach, indicating that the snow came down so fast that they couldn't even fall over. Could it be that frozen comets impacted near the poles or chunks of the frozen canopy hit the poles? Maybe.

The tectonic plates were traveling at around 1–2 mph, and water was still bursting out from under the surface from the deep caverns of the earth as the tectonic plates forced the water out. The tectonic plates were seemingly hydroplaning on the surface of the deep caverns of the fountains of the deep. The pressure from the tectonic plates is what forced the water to burst out from the deep caverns like steady gigantic fountains around the globe at locations called fault lines. And Yellowstone geysers are another example of fountains that burst, but on a much smaller scale currently. The water bursting out from the deep caverns was fresh water; however, the water raining down from the canopy was very salty water. The oceans were rising chaotically, and they were filled with massive amounts of sediment, a mixture of soil, debris, vegetation, and biomass. The floodwater was filled with so much sediment that the turbidity of the water was very high and the translucency was an inch or two. This was about the fifth week of the chaos.

The ark that Noah and his sons had built over the preceding 120 years was built for survival, like a barge; it was not meant to travel, but just to float. Noah and his sons built according to the specifications by God for a large floating barge, with enough storage capacity to safely house two of every kind of creature on Earth, including young dinosaurs and carnivores. Noah placed multiple 9-foot-tall, 2-ton weights along both sides of the ark that countered the swaying, and this stabilized the ark during this chaotic global catastrophe.

The last week of the chaotic flood was much calmer than than the preceding weeks. Volcanic activity had declined significantly. Every island that was formed from volcanic activity and every mountain was covered by 22.5 feet of water all across the globe (Gen. 7:19–20). The fountains of the deep caverns were almost exhausted. The tectonic plate movements that once hydroplaned on the water of the deep caverns below them were now encountering bedrock below, and friction was slowing the movements of the tectonic plates. However, inertia would continue to drive the plates to eventually form deep valleys and high mountains (Psalm 104:5–9). The global temperature was steadily dropping so that the first emergence of ice had formed at both poles. The last Hawaiian island was still being formed under the ocean. The rains from the canopy of water was almost exhausted as well. The entire surface of earth was covered by water. The creatures that once had the breath of life in them were all dead, except for those on the ark (Gen. 7:21–22). Much of the vegetation that had once covered the surface of the earth was covered by the initial asteroid impact and debris. The mixture of debris and vegetation would then begin the processes of coalification, fossilization, and petrification. The massive number of life forms that were covered up in debris would be involved in the process of converting biomass and vegetation into fossil fuels. This was the last week of the catastrophic global flood.

The 40th day:

The 40th day of the Flood had been reached. The fountains of the deep caverns had ceased. The rain had ceased. The Indian tectonic plate had crashed into the Asian plate, and the Himalayas had started to rise. The Andes mountains and all the other mountain ranges formed by tectonic plate movement had started to rise, yet they had a ways to go. All the mountains were still covered by water 22.5 feet deep (Gen. 7:19). The tectonic plates that once hydroplaned slowed their velocity, yet their inertia would eventually cause the great mountain ranges we see today. The waters prevailed upon the earth for the next 150 days. The rain from the canopy of water had finished descending upon the earth and was gone. Global temperatures continued to drop with the entire planet surface covered in water. The ice at the poles continued to expand in glacier tonnage, leaving piles of salt at the bottom of the water below. The volcanic activity had stopped, and the massive amounts of soil and water that covered the volcanoes acted like a cap to cork and cover the magma vents. All seemed calm for the first time in 40 days, except for a large global tide that swept across the waters twice daily. The massive tide reduced static friction of the sediments and made it easier for soil to settle according to density, forming layers according to their kind based on the laws of physics. The only living creatures on Earth were the survivors on Noah's Ark.

Once the 40 days of rain ceased, the waters prevailed (covered) over all mountains on the earth for 150 days (Gen. 7:24, 8:3). The waters continued receding as the polar caps continued to grow in ice and the valleys sank down, and the mountains rose from the tectonic plates colliding, and the inertia buckled the crust of Earth (Psalm 104:5–9). A global wind began for the first time as a result of the polar ice cap formations and the heat from the sun at the equator (Gen. 8:1). After the waters prevailed for 150 days, the first mountaintops appeared above the flood waters. All the while, the soil, vegetation, and biomass that were mixed in the waters were still settling according to density and forming layers of sediment. Due to the violent turbulence of the waters from the catastrophic activity, there were areas around the globe that had more bones, teeth, coral, and shell fragments pooled together and areas around the globe that had sand predominately pooled together. This caused some areas to have more limestone formations and other areas to have more sandstone formations, though both areas were still layered. The waters continued to recede steadily.

Fifth month from the start of flood:

In the seventh month (of the year calendar, which is the fifth month from the start of the Flood), on the seventeenth day, the ark, with its deep berth, rested upon the mountains of Ararat (Gen. 8:4), but the tops of the mountains were still not visible because of the waters.

Seven and half months from start of flood:

It was not until three months later that the mountaintops became visible for the first time since the Flood had covered the entire earth (Gen. 8:5). Noah waited 40 days after the mountaintops were visible to open the window of the ark to send out a raven, but the raven flew to and fro and returned to Noah empty-handed (Gen. 8:6–7). The water steadily receded as the glacial age continued to grow, and the valleys sank down and the mountains continued rising (Psalm 104:5–9). The mighty Himalaya mountain range had formed. Then Noah sent out a dove, but she also returned to him as there was no resting place for her feet. Noah waited another week and sent out the dove again. This time the dove returned with a freshly picked olive leaf in her beak (Gen 8:11). Noah waited another seven days and sent out a dove again, but this time she didn't return.

<u>10.5 months from the start of the Flood (Gen. 8:12–13)</u>:

The ground was dry, and the glacial age had begun. Ocean levels were 200 feet below our current levels because the massive polar ice caps had absorbed vast amounts of water and extended from the poles to the middle of the Southern and Northern Hemispheres. This lowered the flood level to below the continental shelf; thus, England was connected to the mainland, and Australia was connected to Asia via the shelf, allowing the less aggressive animals to flee away from predators. Some migrated all the way from the mountains of Ararat (Turkey) to Australia via the exposed continental shelf. At the one year and 10 day mark, God instructed Noah to leave the ark and take every creature out of the ark so that they could breed and abundantly fill the earth (Gen. 7:11–8:14).

The picture is of the fossil remains of a wooden vessel that meets all the standards described in Genesis for the ark, and matches the location of the resting place recorded in the Bible, thus we may infer that this is Noah's Ark. Points A, B, C, and D outline the vessel with a collapsed roof. *Photo credit: http://www.viewzone.com/noahx.html.*

Shortly after leaving the ark, Noah saw clouds for the first time and saw rainfall off in the distance that was refracting sunlight. Noah saw his first rainbow (Gen. 9:11–16), as God declared the rainbow to be a covenant not to destroy the earth by water again. Animals freely roamed about and were able to travel from continent to continent along particular travel routes called the continental shelf and/or ice. All the continents remained connected because the floodwater receded and allowed life to venture from the mountains of Ararat to places around the globe. With much of the water stored in the ice caps, the ocean levels would remain below the continental shelf level until shortly after the Tower of Babel, during the life of a man named Peleg (Gen. 10:25). His name means "in his days the land was divided." The Tower of Babel existed at a time when all the people had one language and one purpose to become like God. God confused their languages and scattered them around the world (Gen. 10:25–11:9). They traveled to distant continents via the exposed continental shelves and the frozen landscape. After arriving to their new locations, the oceans continued rising as a result of the melted glaciers, and this caused the oceans to rise above the continental shelf. This made Australia an island and protected such animals as kangaroos from the predators of Asia and cut off traveling to different continents until the Vikings and then Christopher Columbus in 1492.

Review: The great and terrible flood was not just water gently falling from heaven. Asteroids impacted the earth, Pangaea broke apart, and the tectonic plates hydroplaned across the surface of deep caverns, forcing water to burst out with soil and multiple volcanic activity on a daily basis. The waters were filled with soil, vegetation, and biomass. The waters were turbulent, causing swirls and pools. The planet was on the verge of being overheated, and then it plunged into the ice age.

Today, we can't perceive the abundance of animal life that was on the earth before the Flood. But to give you an image of how abundant and dense the population was, consider that 15-foot-tall camels have been found in the Arctic Circle along with the remains of woolly mammoths. Imagine a herd of 150 million woolly mammoths roaming. And you know that those mammoths could only survive and flourish with abundant vegetation. I think humans can scarcely take it in, the creative abundance of what God established as far as vegetation and animals. Not only have researchers found woolly mammoths with tropical vegetation in their mouths, but also frozen in the upright position. This indicates an almost instant freeze, with the animals freezing within four to five hours, suggesting that a frozen comet or a chunk of the frozen canopy may have struck the area and sent it into a deep freeze. Also, they have found frozen bobcats, camels, and bison along with the mammoths in the Arctic Circle. This means there were lush green grasslands in the Arctic Circle before the Flood, and it got really cold really fast. If the entire earth was covered with water, then where did all the water go? Some went to the polar ice caps, some got absorbed by the soil, but most formed the oceans. Psalm 104:5–9:

> He established the earth upon its foundations . . . You covered it with the deep as with a garment [*deep* equals ocean waters]. The waters were standing above the mountains. At your rebuke they fled, at the sound of Your thunder they hurried away. The <u>mountains rose</u>; the <u>valleys sank down</u> to the place which You established for them. You set a boundary that they [the Flood waters] may not pass over, so that they [the Flood waters] will not return to cover the earth.

This lets us know that the mountains we see today are much higher than the mountains before the Flood, and the valleys we see today are much deeper. This verse also explains how mountains around the globe have seashells and petrified giant clams (in the closed position) on their peaks.

How did the aquatic dinosaurs die from the Flood? Potentially, some perished by the rain upon the earth; it had high salt and mineral content, and this increased the salinity of water that marine life forms were used to. Some adapted and survived, and some didn't and died. Another explanation would be that some perished by trauma from asteroids and volcanoes, either by direct contact or increased water temperature or from too much debris in the water or acidity of water near volcanoes. Many that survived the flood became stranded in inland seas that were destined to evaporate to dry land, and thus, they died in a massive confined grave as they pooled together to an ever tighter and tighter population as the water receded. This explains how some aquatic marine dinosaurs were killed, as they were unable to adapt to the changes from mildly salty seas to higher ocean water salinity. But some survived the Flood, such as the Leviathan that God references in Job 41. And the Bible records that there were seas from the beginning, so potentially the seas had the same alkalinity, and thus many sea creatures survived the Flood.

There is a redemptive plan by God for all of His creation, and there is hope and a promise that the earth and life will return to Utopia, to Eden. This is found at the end of the book of Revelation and throughout the Old Testament.

Summary: The Flood destroyed life on earth (but two of every kind survived on the ark) and changed the conditions. Knowing what the Flood did and knowing where the source of the waters came from tells us what life was like before the Flood and what the conditions of the environment were like. Those conditions allowed life to thrive. Dinosaurs grew large and humans lived 900+ years. The flood took away the buoyant force of the canopy of water, decreased atmosphere pressure, decreased oxygen concentration, and increased gravity. Consequently, living beings no longer thrive as they used to before the Flood.

<u>Group Discussion:</u> Ponder this: As it was in Noah's day, so it will be when Jesus returns (Luke 17).

1. Few were saved in Noah's day, few are saved today (Matt. 7:13–14). How does this influence you?

2. We know Revelation will be catastrophic, so how does that shape your view of God's judgment with the Flood? Were you under the impression that the Flood was clear waters from a gentle rain?

Chapter 10
No Deserts before the Flood

The title of this chapter seems to be pretty bold. Were there really no deserts from the origin of life until the Flood? Correct. Let's look at some evidence to lay a foundation, and then we'll go to the Scriptures to confirm it. [A satellite image of sand being blown from the Sahara Desert into the Atlantic Ocean.]

The Sahara Desert is located in the northern portion of the continent of Africa. Winds blow the tan sand from the Sahara Desert into the Atlantic Ocean. Scientists observing the sand from the Sahara Desert blowing into the Atlantic Ocean got the bright idea to look into the soil of the ocean floor to determine if there was any evidence that would suggest when the Sahara Desert formed. How? The Sahara Desert sand has distinct reddish-tan granules. Tropical forest soil blown into the ocean has a distinct brownish-green color. Since the wind is blowing sand into the Atlantic Ocean and that sand has a particular color, then by viewing ocean core samples, scientists can detect the color changes to determine when the sand began being blown into the ocean. Then they can discern how long the Sahara region has been a desert, essentially confirming when the Sahara was formed. *Photo credit: Space, Science, and Engineering Center, Madison Wisconsin University.*

When geologists viewed the core samples in their lab, they saw tan sand sediment with striations. Each horizontal striation indicated durations of time or years in the past. With deeper Atlantic floor samples, they saw a distinct color change from reddish-tan sand to darker brownish-green soil. This color change tells the story of when the Sahara region stopped blowing darker brownish-green colored soil into the Atlantic Ocean and began blowing light-colored reddish-tan sand into it. This shift in color is unmistakable and reveals the history of when the Sahara region stopped being a tropical forest and began being a desert.

When Geologist Peter deMenocal of The Earth Institute at Columbia University studied the Atlantic core samples, he estimated that the Sahara Desert has been in existence for an estimated ~5,000 years. This was right around the time of the global flood of Gen. 7—and not millions of years. How did deMenocal come up with this approximate date? He observed horizontal lines in the core sample and estimates of the current rate of soil deposit per year. He then measured the depth of the layer and divided the current rate of soil deposit by distance. This came out to an average of one-fourth inch per 200 years (±). If the rate of deposit has been consistent over the past five millennia, then this dating method is accurate. If there were periods of change in the rate of deposition, then the dates would increase in the margin of error percentage. This method of dating is similar to how evolutionary geologists determine the age of the layers of earth. However, it is difficult to measure accelerated deposits from floods with high soil content, which is exactly what the Biblical flood wrought.

The Sahara Desert has left a sign for all to see to prove that its origin began shortly after the Food of Gen. 7. In fact, the core samples reveal that prior to the lighter reddish-tan dust blowing in from the Sahara desert, there was brownish-green soil blown into the Atlantic ocean. This brownish-green-colored soil represents that the Sahara region was once dense in tropical vegetation, which is exactly how the Bible described botanical life at origins, as abundantly vast.

Peter deMonecal further states that the change from lush vegetation to desert would have taken about 100 to 200 years. However, it seems that the change from the Sahara being covered with tropical vegetation before the Flood and then becoming desert after the Flood may have taken less time than one would think. Vegetation needs water, so to go from an abundant and consistent water supply from the mist rising from the ground to severe drought as it is today wouldn't take long. I contend that Peter deMonecal's numbers could be high and that the transformation was within one to five years after the Flood. Before the Flood, there was tropical forest, with dense lush vegetation. Then comes the Flood saga that kills all the vegetation and then covers it with soil and water. Then the floodwater started to recede as the polar ice caps grew by absorbing the water. While the floodwater receded, what was left behind was soil void of vegetation, void of any plant life that was alive, with only seeds remaining. Those seeds sprouted when the water drained off and when sunlight was able to provide the energy for photosynthesis. The sprouts leaped forth from the ground, yet there was no water to sustain them, the ground becoming ever more dry. The new sprouts died off quickly in weeks, and what was left was the sedimentary sand that we still see today.

The first mention of heat, cold, summer, and winter is not until after the Flood in Gen. 8:22: "While the earth remains, seedtime and harvest, and cold and heat, and summer and winter, and day and night shall not cease." Before the Flood there was no freezing cold, no hot heat, no summer, and no winter.

In addition, the Bible indicates in Gen. 8:1 when the Sahara desert first started having its sand blown into the Atlantic ocean with: "God caused a wind to pass over the earth." This reveals for the first time that there was enough heat at the equator and cold at the poles to create enough change in global temperatures to cause wind. As hot air rose from being less dense around the equator and cold air sank from being denser (high pressure) around the poles, wind, in the form of the jet stream, began for the first time.

Gen. 9:8–17 is the very first mention of the word *cloud* in the Bible. Why is this the first time in the Bible for cloud formations? Well for one thing, before the Flood, there was no rain on the earth because there was no need for rain. Why? "A mist rose up from the ground to water the whole earth" (Gen. 2:5–6). Clouds couldn't form before the Flood because the canopy of water created levels of atmospheric pressure that were too high to allow clouds to form. How does a mist rise up from the earth? it's simple. At night when the temperature drops, the air molecules contract and get smaller with the cooler temperature, and therefore, the air is unable to hold as much moisture. Thus, the moisture in the air settles to the ground in order to bring a homeostasis of the air with the temperature and the moisture. We call the result of water on the ground "dew," and we see it in the mornings on the grass and surface of things on the earth. But Moses, writing under the divine inspiration of the Holy Spirit, chose the phrase "A mist used to rise from the earth and water the whole surface of the ground." Notice that Moses used the past tense to describe this daily event of watering the whole surface of earth. Why? Moses wrote Genesis after the Flood, and therefore this mist that used to rise to water the whole earth was gone. We see a semblance of what used to be when we see the dew in the morning that settles on the grass.

Review: Before the Flood came and destroyed all life, the Sahara region was watered by a mist that rose from the ground. Super-high atmospheric pressure resulting from the canopy of water prevented cloud formation. After the Flood, the highly pressurized atmosphere was reduced to its current condition, which allows cloud formations and rain.

Samples taken from air bubbles trapped deep within glaciers cores and air bubbles trapped in amber indicate that the earth had no deserts and no polar ice caps and that the world was covered in vegetation. How do the bubbles in glaciers and amber tell us this? Well, this takes several steps to get to the conclusion, but vegetation produces oxygen, and it has been affirmed that oxygen concentration in the atmosphere was 31% in the distant past and 50% higher than today's current value. It's the vegetation that produces this high oxygen concentration. Therefore, one could conclude that the earth's vegetation was abundant all around the earth prior to the Flood, and this includes the desert regions and the polar regions. And we could also argue that after the Flood, with the equator now exposed to the harsh sun, then vegetation declined and finally gave way to deserts that were not hospitable to vegetation growth.

High oxygen concentration allows for accelerated growth and the longevity of life. This explains why human beings lived 900+ years before the Flood. And after the global flood of Gen. 7, the number of years lived by Noah's heirs kept declining by almost 100 years each generation. As the oxygen concentration continuously dropped, all living creatures—based on existing DNA code—adapted to the progressively lower oxygen concentrations, which correlated to reduced life expectancy. This includes dinosaurs as well.

How do deserts play a role in reducing oxygen? With the intense heat and lack of water, vegetation doesn't grow well. Therefore, the deserts play a role in keeping oxygen levels below the pre-Flood 31% level, a level at which life thrives by growing faster, stronger, and bigger.

Review: Deserts play a role in preventing vegetation growth and thereby are involved in the reduction of oxygen production. Since oxygen causes life to thrive, deserts indirectly hinder the quality and longevity of life.

The Sahara Desert has remnants of bygone freshwater lakes, with fossilized freshwater aquatic life and residual freshwater deep-bottom algae deposits. This reveals that before the desert came, there were freshwater lakes in the Sahara region that were teeming with life alongside the lush tropical vegetation. NASA has photographs of where the freshwater lakes were and estimates of their size.

In addition, fossils of saltwater marine life and saltwater seashells have been found on top of the freshwater fossil layers. Therefore, before the Sahara was a desert, it was covered with ocean water. This seems odd to have indications of both freshwater lakes and life with evidence of salty oceanic water and life on top of them. How did the ocean water get there?

The global flood of Genesis solves this mystery. Before the Flood, the Sahara region was tropical with lush vegetation and freshwater lakes. Then the salty rain came upon the earth for 40 days and 40 nights, along with water bursting out of the earth from deep caverns, and ocean waters covered the Sahara region for 150 days and then receded over the next six months to complete the nearly one-year flood saga, killing everything and altering the conditions of the Sahara ever since.

The Sahara Desert has remnant fossilized vegetation under the marine life. Therefore, prior to the ocean water and prior to the marine fossils and seashells, there is evidence of vegetation that once thrived in northern Africa. Therefore, Africa once thrived as a tropical forest with freshwater lakes, then it was covered in salt water, and then it became the desert we see today. What changed? Why did northern Africa, once covered in lush vegetation and then covered in salt water, become dry, arid, and the largest desert on Earth (excluding Antarctica, which is considered a desert)? Something drastic that was global occurred around that time period. What major event in the history of the globe occurred around that time period and explains this drastic shift in the conditions of the continent of Africa? The answer: the global flood of Gen. 7.

Scientists have discerned that the Sahara became a desert some ~5,000 years ago. And other scientists doing other research, have found saltwater marine life in the Sahara Desert, and then they have dug a little deeper to find freshwater fossils and remnants of tropical forests. And this is consistent with Biblical events. But then it is mind-boggling to read that they say that Africa was covered by an ocean some 250 million years ago. Well, wait a minute, that's an impossible guess by some evolutionary scientists because the Atlantic Ocean floor isn't lying when its core samples reveal that the

Sahara was lush tropical vegetation just ~5,000 years ago—and above the tropical sediment on land, there are saltwater marine fossils in the desert sand. You can't have the ocean existing 250 million years ago on top of the tropical forest remnants that existed 5,000 years ago. The observable evidence on the floor of the Atlantic Ocean suggests that if the ocean existed and its oceanic marine life was deposited on top and covered up the tropical forest, then the ocean covering Africa by default came after the tropical forests existed—some ~5,000 years ago.

Review: Soil samples in North Africa confirm Peter deMonocal's research that suggests that Africa was once covered with lush tropical forests and with freshwater lakes and freshwater aquatic life. Then, ocean water covered everything and left saltwater seashells and sedimentary layers on top of the lush vegetation and freshwater aquatic life. And Atlantic Ocean core samples proves this event occurred ~5,000 years ago.

Peter deMenocal's time estimate puts us near the time of the Bible's global flood. Some scholars estimate that the Genesis flood occurred ~4,400 years ago (~2,400 BC). And some secular researchers, such as Peter deMenocal, say the Sahara Desert started around 5,000–5,500 years ago. Given that we're talking about two separate interest groups, then a difference of 600 to 1,100 years is a small rate of error: 0.000011%. I've read evolutionists say the date of such and such was about 500 million years ago, give or take a 100 million years (error rate of 20%).

    Considering that the Sahara desert is growing and is the largest desert on Earth, then we may infer that it is the first desert on the planet. And since it is only ~5,000 years old, then it is evidence that before its existence, there was no deserts on the planet. This suggests that something protected the earth at the equator from the sun's harmful rays—that is the canopy of water. Consider that the oldest tree on Earth is roughly ~6,000 years old (some trees have produced double rings, which accounts for claims of trees that appear older); perhaps those trees are from the time of creation. Gen. 7 doesn't record that all trees died (just everything that moved), so it's possible that a tree as old as creation still survives. Notwithstanding, it still supports a young earth perspective, with the point being that there is no tree that is 100,000 years old. And scientists estimate that the Great Barrier Reef is only 4,200 years old; this also supports a young earth and indicates that ocean water may be a relatively new phenomenon. Now, I heard one evolutionist contend that the current growth rate of coral doesn't account for the breadth of coral in the coral reef with a young earth hypothesis. However, this criticism is flawed because growth rates before the Flood were greater than after the Flood.

    We know when the deserts began, but do we know how and why and what caused the Sahara to be a tropical forest? For this answer, we go back to Gen. 1:6. The waters below became the seas and the waters above the expanse stayed there until the Gen. 7 global flood. That water that was above the expanse became the canopy of water. This made everything lighter on the planet, as the heavier atmosphere created a buoyant force that countered gravity by 13.5% to 22%, thus raising the water table in the soil. And when the evaporation of water occurred at such a high atmosphere pressure, it prevented clouds from forming. Then, "A mist rose up from the ground to water the whole earth" (Gen. 2:5–6). This is how the Sahara was able to be have lush tropical vegetation; the canopy of water shielding the harsh sun created ambient global temperatures, and a mist rose from the ground and watered the whole earth.

Review: The Sahara was able to have lush vegetation because God formed a canopy of salt water to shield the earth from the destruction of the sun, yet it allowed the sun's beneficial light through and created an environment where a mist rose from the ground and watered the whole surface of the earth.

The Bible describes that after God had created everything, He looked over all that He had created and declared it was very good (Gen. 1:31). Then on the seventh day of creation, He rested (to make the day holy to the Lord) and declared, "Thus the Heavens and the earth were completed, and all their hosts. By the seventh day God completed His work which He had done" (Genesis 2:1–2). This is to say that God's work was complete, lacking nothing, and very good.

When looking at the book of Genesis, we get a clear picture of the masterpiece, the perfection, and the symbiotic harmony of all creation.

Taking in a bird's-eye view of the Bible, we get even more clarity of what "very good" and "complete" means regarding God's work of creation. A bird's-eye view of the Bible offers a storied history lesson of life and death. With God, there is life. Without God, there is death. We learn that death did not enter all of creation until the fall of Adam and Eve. Therefore, before the sin of Adam and Eve, there was no death anywhere. Since the desert is a symbol of death and represents the suffering of creation groaning for freedom from sin and death resulting from sin, then before sin entered the world, there was no death and by default, no deserts. There were lush tropical forests, and that is exactly what the Atlantic Ocean floor samples reveal and the fossil record in the desert.

Romans 8:20–22 elucidates the point:

> For the creation was subjected to futility, not willingly, but because of Him who subjected it, in hope that the creation itself also will be set free from its slavery to corruption into the freedom of the glory of the children of God. For we know that the whole creation groans and suffers the pains . . . waiting eagerly for redemption.

This means that there will be a redemptive restoration of creation, a time when there will be no deserts again. Utilizing a literal interpretation of the Biblical genealogies, Adam and Eve lived on the earth roughly 6,000 years ago. This fits perfectly with findings that the Sahara Desert didn't begin until some 4,500 years ago at a time shortly after the Flood.

The loss of the canopy to 40 days and nights of rain caused a reduction in atmospheric pressure. All life forms had to adapt to the sudden increase in net gravity, a gradual reduction in oxygen, and the reduction of atmospheric pressure; life forms were able to adapt because of the information already in their DNA, and they lived shorter and shorter lives and became smaller and smaller with each generation on the earth. Prior to the Flood, the canopy hovering above the atmosphere put more pressure on the inhabitants of the earth, which caused the buoyancy effect that reduced gravity by approximately 13.5% to 21.5%. The effect would be approximately 20% of the buoyancy effect one experiences today by swimming in a pool.

Coincidentally, as the dinosaurs had to adapt to the new conditions on the planet after the Flood, so too did all other life forms. Even humans had to adapt. This is manifested in the significantly reduced longevity of life after the Flood. And at the same time that dinosaurs and humans were adapting to the change in the environment, so too did the earth's terrain. Some regions couldn't sustain life and became deserts. Some animal species couldn't adapt, and they too became extinct. But life perseveres and presses onward.

Now we get to the important part. We see the evidence of when the Sahara desert began. We have a plausible interpretation of this evidence that aligns with science. But does our interpretation of the evidence align in harmony with the Bible? This is critical. When anyone is interpreting past or future events and trying to determine if there interpretation is correct, they must always go to the Bible to determine if their interpretation matches up with Scripture. Never should anyone try to force Scripture to fit their interpretations. Why?

When one has their preconceived interpretation and then tries to fit Scripture to their interpretation, they are loving their own ways, their own self, and their own mind. This is idolatry.

What? Come on. Idolatry is worshiping another god, such as a golden cafe. Well, worshiping another god is indeed idolatry. But consider Colossians 3:5: "Consider the members of your earthly body as dead to immorality, impurity, passion, evil desire, and greed, which amounts to idolatry." This verse describes the improper use of love, which is the love of selfish desire/thoughts/actions in this instance. This is putting oneself as a god over their life. And that is idolatry. Idolatry is giving worship/credit to anything that rightfully belongs to God alone. We are to love what the Scriptures say, no matter how foolish one appears in following them. Those that try to fit the Scriptures to their own interpretations fall into the category of self-love: idolatry.

So let's head to the Scriptures to see if our interpretation of the evidence is in harmony with the Word of God. When God spoke the earth and seas into existence, as well as all the vegetation of grass, plants, and trees, God did it completely. How do we know this? Gen. 1:11 reads, "And it was so." Everything God does, all of His actions are in total, complete, without flaw, and without error. After creating the heavens and the earth and all that is visible and invisible, God wrote through Moses, "God completed His work which He had done" (Gen. 2:1–2). God's work of creation was complete, perfect, and lacking nothing.

When God in the flesh, Jesus, was hanging on the cross, He said after six hours of work, "It is finished" (John 19:30). God always does perfect, complete work. When the Almighty God, the slain lamb for the sins of the world, comes to reign in Jerusalem, He will say, "It is done" (Rev 21:6). All three major events by God are written of in the same manner: complete, finished, and done. Nothing is left undone, nothing still needs finishing, and nothing is incomplete.

Therefore, when God spoke, "Let there be" as He created the earth, the seas, the grass, plants, and trees, and when God's Word declared "And it was so," you know that God performed such tasks perfectly. Furthermore, God personally gave an oath testifying to this by saying, "God saw all that He had made, and behold, it was very good. and there was evening and there was morning, the sixth day." After the work God had done on days one through six, at the end of the sixth day as the sun was setting and getting ready to start the seventh day, God looked over all that He had made. And He saw that it was very good. Why? Everything had life, and there was no death, no corruption. Everything was pure, sinless, and perfect.

Therefore, the Bible has implicit testimony from God Himself that there were no deserts in His creation at this time. When God looks over all that He created, He doesn't see the death of the desert, He doesn't see the bareness of the desert, He doesn't see the futility of the desert, and He doesn't see the suffering in the desert. No, quite to the contrary, He sees the lush vegetation teeming with life.

God had created a physical boundary to block and prevent the death of the desert and to block and prevent the death of the polar ice caps. What was the physical boundary? The canopy of water that hovered above the atmosphere and created ambient global temperatures that prevented both deserts and polar ice caps from forming. All deserts are an indication of death resulting from sin. As a result of sin, all of creation suffers futility and is in bondage to the corruption of sin (Romans 8:18–22). Deserts serve a spiritual purpose to remind us that sin is death. And as a result of sin, what was a completely lush planet became flawed.

Review: The Bible is explicitly clear that all that God made was filled with life, and it was very good. The deserts are a symbol of futility, suffering, death, and no life. Therefore, our interpretation that the Sahara Desert and all deserts began after the fall of mankind and after the Flood is in harmony with Scripture.

Deserts, however, are not how the planet will be after Christ comes. Deserts will become a thing of the past. The whole of creation will be reborn, created anew. This topic deserve a whole book, but to give you an overview, I refer to Isaiah 24:18b–23: "The windows above are opened, and the foundations of the earth shake." This is in parallel with Revelation 6:12–17, as it says that "a great earthquake . . . and the stars of the sky fell to the earth." This is saying that massive meteors and asteroids will impact the earth in the future.

Now, let's go back to Isaiah 24:19: "The earth is broken asunder, the earth is split through, the earth is shaken violently, the earth reels to and fro like a drunkard and it totters like a shack." Now, let's review Revelation 6:14, which says that "every mountain and island were moved out of their places." The verses seem to say that asteroids will hit the earth and a series of impacts will set in motion the literal splitting of the earth. There is a likely place on the planet that seems like a zipper, or a seam of the earth. It's in the Atlantic Ocean and runs the entire length of the ocean floor to Iceland. This is where the mantle of the earth pushes up and pushes apart the American tectonic plates from the European and African tectonic plates.

When the earth splits asunder, all the earth's oceans will pour into the superheated mantle and into the core. This will cause water to explode up and shoot outside our atmosphere like a massive Yellowstone geyser. This will reconstitute the canopy of salt water back around our atmosphere. And this also will increase atmospheric pressure, which will recreate the buoyancy effect. This buoyancy effect will reduce the effects of gravity. The recreation of the canopy and the buoyancy effect will be a restoration of the Garden of Eden environment. The canopy will create ambient global temperatures. And there will be no polar ice caps and no deserts. And replacing those barren regions will be lush plant life. With the loss of oceans, lush vegetation will grow where the oceans once were. This will increase oxygen concentrations back to pre-Flood conditions. As all the ocean water hits the superheated core and shoots up into space, this will act like a huge jet engine and cause the earth to begin spinning faster again. This will stop the earth from wobbling like a drunkard or a slow spinning top from the multiple asteroid impacts and cause the earth to spin smoothly like a top again. The increased spin will mean that the days will be about 17 hours long, as they were at creation. This explains why Revelation 21:1 prophecies that "there is no longer any sea."

This explains how humans again will live to 900+ years again. Isaiah 65:17–26:

> For behold, I create new heavens and new earth . . . No longer will there be in it an infant who lives but a few days, or an old man who does not live out his days; the youth will die at the age of one hundred and the one who does not reach the age of one hundred will be thought accursed . . . as the lifetime of a tree, so will be the days of My people . . . the wolf and the lamb will graze together.

This is just how life was in Gen. 1:30–31: "To every beast of the earth and to every bird of the sky and to every thing that moves on the earth which has life, I have given every green plant for food . . . God saw all that He had made, and behold, it was very good." The result of the earth splitting asunder from God's wrath is that conditions on the earth will return to the pre-fallen state, which means no deserts.

One thing that I've always wondered about is what life would be like with no thorns; imagine a rose bush with no thorns. I haven't heard of anyone growing rose bushes in increased oxygen and increased atmospheric pressure—which are pre-Flood and post end-times judgment earth conditions—but I hypothesis that the thorns would disappear with increased pressure and increased oxygen concentration. The deserts we see today will be gone one day soon, when Christ comes and destroys all evil. There will be a new heaven and a new earth.

Chapter summary: Before the Flood, there were no deserts. There were tropical forests, lush vegetation, and freshwater lakes. Those tropical forests left sediment deposits on the Atlantic Ocean floor and on the land of the Sahara region. The remnants of tropical forests and freshwater aquatic fossils were covered up by oceanic fossils and sediment from the Flood. The

desertification process probably took one year from the first rains of the Flood till the waters receded. After the Flood, there was no longer any canopy of water to shield the equator from the destructive sun, no deep caverns of water, and no mist that rose from the ground and watered the whole earth. This caused the deserts. The deserts limited vegetation growth, which limited oxygen concentration, which limited life from thriving.

Group Discussion:

1. The brief history of the deserts on Earth authenticates the Bible and a young Earth hypothesis. If you are an old earth theist, does this influence your view? If you are a young earth creationist, does this strengthen your faith in the Bible and God?

2. Ecclesiastes 1:9 says, "That which has been is that which will be, and that which has been done is that which will be done." Connecting this with the calamity coming as prophesied from Isaiah 24 and Revelation 16 and the restoration of the canopy above the atmosphere, how does this change your perception of what the environment was like at creation?

Chapter 11
When and What Caused the Polar Ice Caps and the Ice Age?

If the glacial age formed slowly over thousands of years to hundreds of years, there would never be hundreds of millions of woolly mammoths (and other creatures) frozen in the Arctic Circle. They would have migrated south for food as the cold crept in and killed off the vegetation. Elephants are descendants of woolly mammoths, and they require ±200 liters of water each day and tons of food, which means that no elephant could live in the current icy Arctic Circle. The evidence that millions of mammoths have been found in the Arctic Circle means that the area was once lush with green vegetation. Digested food moves from the mouth to defecation in about 8 hours. And the fact that some mammoths were found frozen in the standing position with tropical flora undigested still in their digestive tract suggests that the glacial age began quickly, such as with the Flood that could have caused the mammoths to swim to stay alive. Then the waters froze, and the mammoths froze as they tried to swim to safety. Thousands of years later, when the ice began melting and they were discovered by researchers, it seemed clear that some of them had been frozen in a standing position, or they had been frozen while swimming to survive a flood.

The current evolutionary glaciologists' hypothesis of a slow buildup to the glacial age has a major hole in it when one considers the frozen creatures found frozen in the standing position in the Arctic Circle with undigested tropical flora in their digestive tracts. This information gives us a clear picture of what life was like before the Flood at origins. There were billions of animals around the globe roaming the lush green earth. They roamed from around the equator to the poles because there were no polar ice caps, and there was lush vegetation worldwide. This information is in perfect harmony with the Bible.

Deep beneath Antarctica, Russian researchers have discovered entire tropical and subtropical forests along with freshwater lakes. And there are records of drillers in Alaska finding trees that were 300 feet tall and frozen vertically in the bottom section of a 1,000-foot-tall block of ice. Alaskan temperatures in the area where these trees were discovered must have been vastly different before the Flood. The point is that there probably was a quick freeze that occurred in both polar regions during the Genesis flood saga.

Review: Woolly mammoths that were found in an upright position with undigested tropical flora in their digestive tracts indicate that there were no polar ice caps at the time of origins, but there were vast areas covered with lush tropical forests—also, that the glacial age came suddenly, not over thousands of years or hundreds of years. The observable evidence is in harmony with the Bible, but not with a slow build-up hypothesis as evolutionary glaciologists believe.

Evolutionary scientists' tendency to fit billions of years to the age of the earth has forced several conclusions from observations to fit their preconceived paradigms. For example, mankind sees the rate of snow accumulation on the polar ice caps today and surmises that the rate of snow buildup has always been at this rate. But, if the snow buildup today is much slower than the buildup say, 4,300 years ago, then the assumptions of the age of the ice pack would be wildly off and much younger than speculated. Likewise, mankind sees the rate of soil accumulation today and surmises that this rate of accumulation has always been relatively constant because this fits their predetermined belief of an old earth. Thereby, the age of the ice age is determined to be very old, and subsequently so too is the age of the earth. However, this leap of faith is fatally flawed because the observable evidence discussed above suggests that the buildup of ice happened quickly in the past. Thus, the age of the earth is much younger than speculated.

The best method of determining the age of items is to utilize the rate at which items decay on a molecular level. This is done by measuring the loss of electrons circling the nucleus and the atomic mass reduction of a radioactive element, using the radioactive isotopic dating technique. Radioactive elements (uranium, plutonium, etc.) are unstable due to so many electrons circling their nuclei. To become stable, they lose an electron at a known rate until the elements become stable. This known rate of losing an election is called the constant rate of decay. The first stable element on the periodic table after the radioactive elements lose enough electrons to become stable is lead (Pb). The quantity is measured with an ion mass spectrometer. A mathematical formula is implemented involving the number of stable lead ions, the number of unstable radioactive polonium ions, and a constant rate of decay to determine the age of an item. Mankind takes a tremendous leap of faith by suggesting that the rate of decay that we measure and perceive today has always been constant. This is a huge leap of faith, as no one alive today lived thousands or millions years ago to observe the rates of decay back then. And there is some compelling evidence that indicates that the rate of decay can accelerate and has accelerated from time to time with certain traumas on the earth. This means that the age of the earth is not nearly as old as we are being told. And by default, life on earth has been in existence for a much shorter time period.

The basis for this seemingly accurate dating system is radioactive isotopic dating, which uses the multiplier "constant rate of decay (CRD)." Here's an example:

lead (Pb) ions / # polonium (Po) ions X CRD = age of the item tested.

I'm not saying that the calculations are wrong. I'm saying that the multiplier for the formula, the CRD, is not constant and therefore the formula is flawed. And if it's flawed, then all conclusions based on that multiplier being constant will be wrong.

To be fair, if the rate of decay was constant, then their calculations would be correct. And the age of the earth would indeed be billions of years old. But the CRD is not constant. In nature, there are several examples where the rate of decay has accelerated with trauma and that establish strong evidence that the rate of decay is not constant. In fact, the examples will show evidence that the age of the earth is very young and that traumas to the earth have accelerated the aging process drastically.

Nuclear physicists are likely screaming right now saying, "Do you know how many kilojoules of energy are required to ionize uranium 238 to accelerate the decay process?" It would take three to five lightning bolts for each uranium molecule. Well, there are currently 100 lightning bolts every second on the planet. That is 3.15 billion lightning bolts every year. And the global flood of Gen. 7 was so catastrophic, with asteroids, volcanoes, and fast-moving tectonic plates, all of which produced electric discharge for lightning. In fact, volcanoes have been registered to produce 7,000 lightning bolts per hour. Now imagine hundreds of volcanoes erupting daily for the 40 days of the Flood. Additionally, multiply that number by billions more to equal the electric charge generated from all the tectonic plates sliding across the surface of the earth. The amount of electrical discharge that occurred during the Genesis global flood accelerated the aging process. And there are several ways of accelerating the decay of elements. There are alpha particle emissions, beta particle emissions by electron ejection, nuclei capture of an electron, positron ejection, and even nuclei fragmentation into two smaller elements.

There is additional evidence that (a) there was a global flood, and (b) the Flood accelerated the decay rate. This evidence is found in microscopic crystals called zircon. Zircon is often found in black mica (a biotite that is a black, dark brown mineral occurring in many igneous and metamorphic rocks). When U238 decays and loses alpha particles to form Th234, the byproduct is helium, and the helium often gets stored in crystals called zircon. Within zircon are helium concentrations that exceed the normal rates of decay under natural conditions to the tune of 1.5 billion years worth of accelerated decay. In other words, helium dissipates from the zircon crystals at a known rate, and uranium loses alpha particles to form helium at a known rate, so since the concentration of helium in zircon exceeds this rate of equilibrium by 1.5 billion years, then there was an event on the earth that caused a massive acceleration of uranium 238 decay in order to have that much helium stored in zircon crystals in igneous rocks. And since the zircon crystals still have this exceedingly high abundance of helium, then this event occurred recently, within 4,000–8,000 years ago. Otherwise, if this accelerated aging event occurred millions of years ago, then the helium in the zircon crystals would have dissipated by now to an equilibrium state.

Review: Mankind has invented a method of determining the age of things that fits their preconceived notion that the universe is billions of years old. They didn't have a blank canvas; their beliefs determined their interpretations and guided their formulas. Zircon crystals indicate that something traumatic occurred on the planet recently, and it accelerated the aging process. Don't be fooled when you hear that the earth is billions of years old and that the glacial age began millions of years ago.

The destruction of the Flood was so powerful, with so much energy released, that it solves how rates of decay were accelerated. My hypothesis is that asteroids kicked off the event by slamming into the crust of the earth, but along the way, the asteroids collided with the frozen external arch of the canopy of water. The impact of the asteroids fractured the crust of the earth and initiated the breaking apart of Pangaea. Proof of the asteroids' impacts is a layer of iridium beneath the layer of the soil. The breaking apart of Pangaea caused massive and multiple volcanoes to erupt simultaneously around the globe. When volcanoes erupt, they usually have lightning accompanying them. With the tectonic plate movements, this would have been accompanied by massive amounts of static discharge (lightning) in the soil as well. The breaking apart of Pangaea caused the fountains of the deep to burst open, and massive numbers of lightning strikes were the result. Altogether, there was so much energy released that there was plenty of energy to ionize enough U238 to accelerate the aging process. When mankind does radioactive studies today using today's rate of decay, there appears to be billions of years of existence. But it was the destructive global flood that caused billions of years of acceleration of decay. And evidence of this is the amount of helium stored in zircon crystals in minerals in igneous rocks, revealing that there were 1.5 billion years of acceleration from around the time that the Bible records the global flood.

The following are some examples of how the rate of decay is accelerated by processes we can observe and test today. And when this accelerated processes occur, they will still be aged much older than reality.

Petrified trees: We are told by people who weren't there to observe that petrification takes 500,000 to millions of years with a slow process of decay, yet Washington's Mount Saint Helens erupted in 1980 and has produced petrified trees in only 30 years time. These petrified trees are indistinguishable from other petrified trees that are said to have taken 500,000 years to petrify. That's an extreme acceleration of petrification from the believed CRD—1/16,700th of the time. There are WWII materials that have petrified, and that's only 70 years. There are hundreds of cases of petrification occurring in a short amount of time. Of course, it should be said that those sages that tell us it takes 500,000+ years to produce petrified wood were not there to observe this long process; they make an assumption based on the CRD and proclaim it as fact. Others argue that the 1980 eruption caused erosion that unearthed petrified trees from previous eruptions, such as from 1842. Even if an atheist argues that the petrified wood was from 1482, 1,200 BC, or 2,500 BC, the concept is the same; trauma accelerated the aging process and the CRD is none the wiser.

Coalification: We are told that the formation of coal primarily took place 300 million years ago with a slow process of decay. Yet, in a lab, we can mimic the Flood conditions (of a buried piece of wood near a volcano) by taking a piece of wood, adding trace elements of clay, H2O, sealed in a vacuum (minus air), and adding heat (150°C) and time (<u>eight months</u>) and create 100% coal that is indistinguishable from the coal formed naturally by all the techniques so far applied to it. Dr. Robert Gentry has repeatedly observed lead-uranium ratios in coal that demonstrate that "both the initial uranium infiltration and coalification could possibly have occurred recently within the past several thousand years" (*Science*, October issue, 1976). The evidence indicates a severe acceleration of hundreds of millions of years from what evolutionary geologists believe took place long ago. Since the CRD is used as the basis for calculating the age of coal to be hundreds of millions of years old, then this illustrates the flaw in believing the CRD is constant. Therefore, the CRD is only constant in the imagination, is not evidence for an old earth, and means radiometric dating is merely radiometric fiction.

Petroleum: We are told that oil generation was a very slow process that began hundreds of millions of years ago. Yet, algae with water cooked at high temperatures and pressure converts the algae to petroleum in minutes, and chicken byproducts (such as adipose (fat), veins, ligaments, cartilage, and the like) are heated up to 150°C, and 100% petroleum can be produced through distillation techniques in 30 minutes. Since we can produce oil in a brief amount of time that is indistinguishable from oil that allegedly is hundreds of millions of years old, then the CRD that forms the basis for determining the age of oil is radiometric fiction. This illustrates how the CRD is just an aspect of a formula to support an old earth that subsequently adds the necessary time element for evolution, rather than scientific fact.

Fossils: We are told that fossils turn to stone through a slow process. However, no fossil ever forms naturally as sediment is slowly deposited over 100,000 to a million years because vegetation or biomass doesn't wait around before it decays into dust, so there is nothing to fossilize. Only with a quick deposit of soil, moisture, and pressure will the organic material change to stone before decaying into its base elements and being converted to nutrients for soil. Just the fact that we have fossils is observable and testable evidence that there was an acceleration of soil deposited quickly as in weeks to months, that is, as in the global flood. And there is another element that indicates how fast things occurred during the catastrophic Genesis flood: there are dinosaur bones that didn't even have time to turn to dust because a deep freeze came over them in the Arctic Circle and froze them.

Therefore, considering the preponderance of the evidence, the CRD is not constant today and has never been constant. In fact, petrifaction occurs in ~1/16,700 of the time necessary, coalification occurs in ~1/30 millionth of the time necessary, and petrolification occurs in ~1/877 billionth of the time necessary with trauma. To add more weight of evidence that the CRD is not constant, a German lab (Power Spectrum Analysis of Physikalisch-Technishe-Bundesanstalt) released decay-rate data after a 15-year study, confirming that uranium-226 had seasonal and monthly variations. This is observable and testable evidence that the CRD is not constant. Also, lightning strikes 100 times every second around the globe, ionizing elements and accelerating the decay process. The helium stored in zircon crystals is the final nail in the coffin. Not only is the CRD not constant, but it never has been constant. Thus, the purported evolutionary time line of the age of the earth and life on earth is also seriously flawed. Since evolutionists declare that the earth is 4.6 billion years old, one would suspect that if they are wrong and the earth is young, then their calculations that support their view would be off by huge numbers. And that's exactly what we find. The evidence is that the calculations supporting the 4.6 billion years is off so much that it supports the young earth view. Science and young earth creationists don't have a problem with trauma accelerating the process of aging. If fact, it fits right into the Biblical record and the scientific method.

Review: With extreme trauma to the earth, such as the Flood, the rate of decay (aging process) was accelerated. The CRD is not constant. Therefore, radiometric dating—the foundational bedrock for evolution—is radiometric fiction.

What does this have to do with the polar ice caps and the ice age? Well, since nature and mankind can accelerate the rate of decay (aging process), and we know the CRD is not constant, then when evolutionary glaciologists use evolutionary geologist's time lines as foundational to date the ice age, you know their purported dates are in error. And the ice age didn't occur hundreds of millions of years ago, and the glacial age didn't occur 800 million years ago, and there weren't five glacial ages— there was one glacial/ice age that occurred ~4,400 years ago (which ebbs and flows) from the severe cold resulting from the aftermath of the global flood. Furthermore, they are using flawed dating techniques to establish an age that supports billions of years to allow evolution enough time and to promote theories that are contrary to the Bible's time line.

How did the polar ice caps form, when did the polar ice caps form, and when was this ice age? The Bible gives us clues to this mystery. In the chapter "Canopy of Salt Water," we discussed the source of the 40 days and nights of rain. And that source was the canopy and the deep caverns of water. This canopy was formed on the second day of creation, as described in Gen. 1:6–8. Also, we learn from Gen. 7:11 and 8:2 that water burst out of deep caverns for 40 days and nights. Combining these verses together with God forming the seas on the third day of creation, this gives us the notion that water was stored in three locations: the canopy, the deep caverns, and the seas. The canopy of water and the deep caverns are what God used to flood the earth, and the Flood is the cause of the glacial ice age.

How do we know that there were no polar ice caps during the time of Adam and Eve? This canopy initially hovered spherically around the atmosphere with the greater part of the water/frozen arch structure near the equator because of the moon and the spin of the planet. The canopy would have provided ambient global temperatures and shielded the earth from harmful ultraviolet light and extreme variances of hot and cold temperatures globally. After all, Adam and Eve were naked all the time, and after sinning, they only covered their private parts. At the time just prior to the global flood, the global ambient temperatures would have been probably around 75°F at night and 85°F during the day at the equator—and potentially 50°F at night to 70°F during the day at the poles. This is not cold enough to cause ice formation and does not provide enough change in temperatures to create a jet stream. Gentle breezes would be the forecast.

Well, couldn't this mean they lived near the equator? Yes. So let's go further. Prior to the Gen. 7 flood, there couldn't be polar ice caps or deserts because there wasn't enough global change of temperature from the equator to the poles to cause a global wind. We learn that after the Flood in Gen. 8:1 that God causes a wind to pass over all the earth; this is our jet stream, and this is caused by changes in global temperature. Also, from a creation point of view, as was pointed out in the chapter on deserts, God created everything complete, and everything was very good, lacking nothing, and there was no death. The polar ice caps are a symbol of a planet suffering the pains of sin and death. Antarctica is considered to be the largest desert on the planet, a vast waste land of arid cold death, the perfect symbol of the absence of God, who represents light, energy, warmth, and life. Therefore, when God looked over all His creation and inspired Moses to write, "God saw all that He had made, and behold, it was very good" (Gen. 1:31), there was no death, and since deserts and polar ice caps are representatives of death, they were not existing at that time. Romans 8:20–22 reveals that all of creation suffers as a result of sin. One of the judgments of sin came about at the global flood, which initiated the four seasons, the hydrological cycle, and death represented by the polar ice caps and deserts. The four seasons represent the death (fall), burial (winter), resurrection (spring), and life (summer). The four seasons were not created on the fourth day of creation. Either God created the earth and it already had a tilt at the time of creation and the canopy of water prevented the four seasons, or the global flood and large asteroids impacting the earth caused the earth's tilt of 23 degrees on its axis and caused the four seasons.

Therefore, when God created the universe, He did it completely, lacking nothing. There couldn't be deserts and polar ice caps because they represent the near absence of life. And no earth-shattering event occurred to alter the perfect climate God created until the Flood came and changed life on earth.

The Bible implicitly declares that there were no polar ice caps until after the Flood and explicitly declares there was no death prior to the fall of mankind to sin. Therefore, we can infer that prior to the fall of mankind, there were no polar ice caps and no deserts. Both represent barrenness and a lack of life. They are almost completely dead regions, void of life. The polar ice caps and the deserts (including Antarctica) don't come close to what God declared after creating everything. He said, "It was very good".

Review: There couldn't be polar ice caps and deserts prior to sin entering the world because all that God made was very good and complete. And there were no earth-shattering events until the Flood to change what God had made. Also, the deserts and polar ice caps are representatives of what sin has brought, which is the absence of life. And all creation suffers and groans for freedom from the bondage of sin, which is death.

When did the ice age (or the glacial age) occur? The massive ice age that scientists refer to, which allegedly occurred 2.6 million years ago, occurred as a result of the global flood of Gen. 7. How is this possible? Scientists look at glacial cores and observe layer upon layer, and knowing the rate that each layer builds up today, they count backwards and hypothesize that the first of five glacial ages started 800 million years ago, and the most recent ice age started 2.6 million years ago. This is hokum. They are interpreting the data to fit their already conceived time paradigm that attempts to provide enough time for evolution. They weren't there, and there are holes in their hypothesis, which means their hypothesis is faith based. Below is an alternative interpretation that is based on the Bible and is in harmony with the observable evidence. *Image credit: www.theresilientearth.com. by Doug L. Hoffman.*

The global flood of Gen. 7 was caused by massive amounts of asteroids and volcanoes and tectonic plate movements and heat and water and dirt bursting out of deep caverns and rain for 40 days and nights globally. The temperatures would have been very high in the initial period of the Flood, so high that if there wasn't rain to calm things down, the earth would have baked to super-high temperatures, killing everything, even Noah and those inside the Ark. With the rains preventing a global inferno, the temperatures stabilized, though it was very warm. Toward the end of the 40 days, the temperature would no longer be stabilizing and would succumb to supercool temperatures, as the asteroid impacts would have stopped within the first couple of days, and the volcanoes subsided, and then the temperatures started to plummet quickly as a result of the sun being blocked and a globe covered in water. Potentially, this record-setting cold freeze, with such a prolonged period without sunlight, may have plunged the planet from a comfortable 66°F high at the poles before the Flood to a low of 0°F at both poles toward the middle of the Flood, plummeting at the poles to -50°F by the end of the Flood. Having no sun for 40 days and 40 nights globally would have first calmed the high heat from asteroid impacts, tectonic plate movements, and volcanoes eruptions and then plunged the globe into a massive ice age toward the end of the Flood. This would be so severe and extensive that the ice from the North Pole would have reached as far down as central North America (this is illustrated in the drawing), all of Europe, and most of Asia. And from the South Pole, it would have reached as far up as southern South America, South Africa, and the southern portion of Australia. At this time during the glacial age, the ocean levels would be below the continental shelves because the ice absorbed a lot of the water out of the oceans and left behind massive salt mines.

The entire flood saga took almost one year; there was 40 days and 40 nights of rain, 150 days of water prevailing over everything, and then about 6 months of water absorption. Combining this information with the mammoths being frozen alive indicates that the poles began to absorb ice earlier in the Flood saga and progressively grew over the following year. This caused the ocean levels to drop lower and lower as more ice formed until the ocean levels were at the level of the continental shelves. After almost a year from the start of the Flood, the glacial age was at its peak. The ocean levels stayed below the continental shelves until immediately following the time of the Tower of Babel. From the peak of the glacial age (almost one year after the Flood), the glaciers started to recede.

Review: Through deductive reasoning, one can see that the records in the Bible implicitly indicate that the glacial age began within the first year from the start of the Flood. Key indicators include the woolly mammoths that fed on tropical fruit in the Arctic Circle and were frozen alive, 150 days of water prevailing over the entire earth, the global wind that started after the Flood while the waters were receding, and the statement, "Everything God created was complete and very good."

There is no question that we had a glacial age at some point in the past. The extent of the glacial age is observed through the directions of rock scourings and scratches and patterns that are unique to glaciers. There are valleys around the globe that have the typical horizontal cutting and scraping formations associated with glacier movements. This is seen in valleys once a glacier has receded. Rock deposits the size of cars can be carried by the massive force of the glacial movements and deposited in regions 1,000 miles away from their origin. In fact, there are many locations around the globe where the land is still rising from once being compressed by the massive weight of ice. This process is called a post-glacial rebound/glacial isostasy. Since the land is still rebounding from the huge amount of weight of ice being removed, this suggests a recent removal—as in thousands of years as creationists believe, and not millions of years as evolutionists believe. Does the Bible give additional clues about when the Glacial age occurred? Yes. Gen. 10 and 11, the narrative picks up three generations from Noah (King Nimrod was Noah's grandson), all of mankind spoke one language, and they wanted to make a name for themselves, so they built a tower to reach the heavens, and God in the flesh (Jesus) came down and confused their language and scattered them around the globe. After the language was changed to the multiple languages that we hear today, the people migrated away from each other and from the central location of Babel (later known as Babylon and now Turkey). Since this is shortly after the Genesis flood, we may infer that the people migrated away from each other to different continents along the continental shelves or ice; both allowed each continent to be accessible at that time. In order to have the continental shelves exposed as dry land, massive amounts of water had to be frozen as glacial ice to reduce the water level below the continental shelves. And that is what the Bible records—that the waters receded. Therefore, we may conclude that as the glacial ice grew toward the end of the year-long flood saga, the more the floodwater receded. Thus, the formation of the glacial age was instrumental in causing the Genesis flood to recede. *Image credit: www.colonial.net/Plate tectonic theory.*

The image illustrates a continental shelf adjacent to the coastline; currently the shelves are hundreds of feet below the sea level because the ice age has melted, but during the ice age, the sea level was much lower. And during this lower sea level and at the time of the Tower of Babel, mankind was able to travel to different continents via exposed continental shelves and ice that connected continents. After successfully walking to other continents, it was at this time that enough ice melted that the waters of the oceans rose above the continental shelves and cut off each cultural tribe with vast oceans.

This is further Biblical evidence that the glacial age existed during the days of Peleg, who was born 101 years after the Flood (fifth generation from Noah, Peleg means "in his days the earth was divided" Gen. 10:25), ~2250 BC, some ~4,265 years ago.

Review: Pangaea broke apart during the 40-day flood, the glacial age began as the polar ice grew, and the floodwater receded into ice until the water level was below the continental shelves, marking the peak of the glacial age. Peleg was born 101 years after the Flood; during his time is when enough ice had melted to cause the ocean waters to cover the continental shelves that connected each continent and cut off cultures from each other until modern times. This occurred after the scattering of the people from the Tower of Babel event (Gen. 11).

How does the Bible's global flood account for glacial core samples that determine oxygen concentrations were 50% higher in the distant past than today's current values? Higher oxygen concentrations before the Flood is in perfect harmony with the Bible. We know that there was abundantly more vegetation on the surface of the earth producing oxygen and that life thrived in the higher oxygen concentration environment and that human beings lived 900+ years before the Flood. So we are in accord with oxygen being 50% higher before the Flood and being reduced after the Flood with the onset of deserts, polar ice caps, and vast oceans that physically limit vegetation from growing to produce oxygen.

How does the Biblical account of the global flood deal with all the layers of snowpack at the poles? With each day of rain during the Flood, there would be changes of temperatures from the highs of the day to the lows of the night. This change in the apex of warmth and the low of the cold would allow for changes in the type of precipitation deposited. During the beginning of the Flood, there were only raindrops globally because of the high global temperatures from the fires, asteroid impacts, volcanoes, and tectonic plate friction, but toward the end of the 40-day flood, the global temperatures would gradually decrease and decrease. At some point during the 40 days of rain, water raining down from the canopy at the polar regions would have been falling as snow, hail, and sleet at the poles. The temperatures started dropping drastically in the polar regions, and the snow would have fallen most of the day. Presuming one to two hours of sleet and rain, this change of precipitation intermingled with rain and sleet, and freezing floodwater would have caused new layers of ice on a daily basis. Spanning the year-long Flood saga, temperature variations throughout the day of warm and cold would have caused layers of ice deposited on the North Pole and South Pole; combining this with a twice-daily global tide created the layers of ice we see today.

Some historians estimate that the Egyptian dynasties occurred around the estimated time of the Biblical flood, and therefore, this debunks the Flood. The estimates that are given for the dates of Egyptian history are just that, estimates. Biblical theologians, combining the records of Genesis and Exodus with observable evidence of the pyramids, estimate that the Egyptian dynasties began after the

Flood, when one of Noah's sons (Ham) settled near the region. Then, the Tower of Babel occurred (Gen 11). Then, the Egyptian era began after the Tower of Babel, and the empire was already established when Joseph was sold into slavery in Gen. 37. The peak of the dominance of the dynasties of Ancient Egyptian is explained in Gen. 41 and Exodus 2. This was aided by the mass arrival of the Jewish people. The building of the pyramids were probably best performed by the Jews and accelerated by Jewish slaves during their 430 years of captivity (Ex 12:40). Jews are famous for providing high-quality work because of the blessing of God. But the Egyptian dynasties came to an abrupt end with Moses leading an estimated two million people (Ex 12:37) out of Egypt, and Egyptians were compelled by God to give the Jews their silver and gold, crippling the Egyptian economy (Ex 12:35–36), and this led to the immediate downfall of the Egyptian reign.

Don't get caught up in evolutionary geologists and archeologists saying that the Egyptian era debunks the timing of the Flood. Evolutionists are always willing to suggest a date that is incongruous with the Bible. The Bible implicitly explains how the pyramids were built—on the backs of skilled Jewish laborers (Exodus 1–2). There is plenty of time for Egyptian monuments to have been built and dynasties to have reigned from 2300 BC till the induction of Joseph and the arrival of the original 70 Jews in 1875 BC to the great pyramids being built by the slave labor of the Jewish people, to the mass Exodus with Moses around 1445 BC.

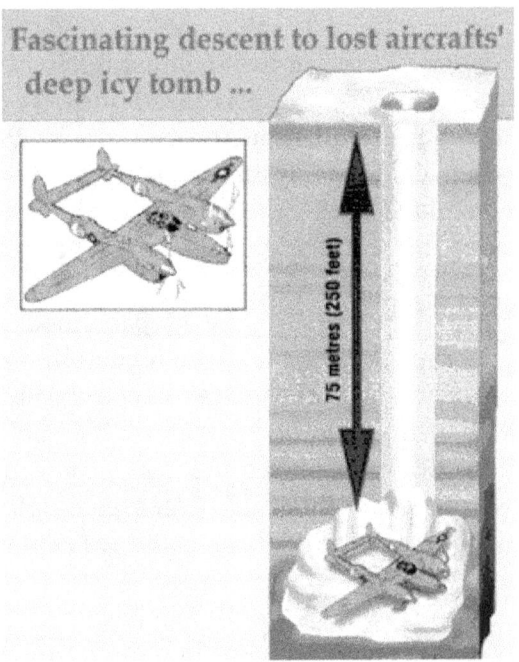

Review: Many layers of ice formed from the global flood, and then the vast glacial age began. Theologians studying Bible records discern from the genealogical records that the Flood occurred around ~2,350 BC—2,400 BC.

Fast-forward to today, and a scientist sees the current rate of deposit of snowfall, measures that amount, and then works backward to determine the age of the glaciers. The scientist then misinterprets the data and sees layers in the glacial cores in terms of millennia instead of hours or days or weeks or months. In fact, there is an odd situation that throws a monkey wrench in evolutionary glaciologists' time line: glacier core samples have layers, and evolutionists count 135,000 annual layers with 10,000 feet of ice built up. That equals an annual deposit rate of 0.07 ft/yr. They don't know that those are yearly layers and not a more frequent layer formation pattern—they are speculating. There is proof that their calculations are wrong. A WW2 airplane was buried 263 feet down below the ice, and there was only 48 years of snowfall between the estimated origin date (1942–1948) to the discovery of the plane. That comes out to (263 feet/48 years) an annual deposit rate of 5.5 ft/yr. Therefore, dividing the estimated 10,000 feet of ice by 5.5 feet per year equals 1,824 years of glacial core sampling, not 135,000 years as evolutionary glaciologists contend. The conclusion is that the layers that evolutionists proclaim to represent yearly deposits are not annual layers. No, they are temperature swing layers from warm to cold, warm to cold, and so on, not annual layers. *Image credit: www.creation.com/the-lost-squadron.* What is the motivating factor for evolutionists to interpret the layers in years versus temperature swings? The answer also relates to why they deviate from the Bible. There are two reasons why this deviation from the Bible occurs:

1. Evolutionary glaciologists use rates of deposit that are seen today and extrapolate the rate of deposition as if they have been relatively constant. This is a voluntary choice to choose dates and

interpret data that support an evolutionary time line and that by default conflict with the Bible. And some rely on man's wisdom as the standard for discerning what is truth, instead of relying on the wisdom of God and what His Word reveals as the standard (Proverbs 3:5–6).

2. The unsaved person is unable to discern the truths in the Bible because their spirit is hostile and alien to God's Spirit. This is an involuntary choice because the Truths in the Bible are spiritually discerned; thus, without the Holy Spirit revealing the truths of creation and the origins of life, they can only be fairy tales (I Corinthians 2:14–16).

Both categories may contain saved people bound for heaven, but their effectiveness in furthering the kingdom of God is reduced because it is harder to logically share the Gospel to someone and tell them to believe in a God that has errors in His Word.

This fulfills two end-time prophecies regarding evolution and creation. Romans 1:18–32:

> That which is known about God is evident within them . . . For since the creation of the world His invisible attributes have been clearly seen, being understood through what has been made, so that they are without excuse. For even though they knew God, they did not honor Him as God or give thanks, but they became futile in their speculations, and their foolish heart was darkened. Professing to be wise, they became fools.

And 2 Timothy 4:3–4 says:

> The time will come when they will not endure sound doctrine; but wanting to have their ears tickled, they will accumulate for themselves teachers in accordance to their own desires, and will turn away their ears from the truth and will turn aside to myths.

Review: Evolutionary glaciologists postulate that there were five glacial ages, with one allegedly occurring 800 million of years ago to coincide with the massive number of years required for evolution, whereas a believer in God's word matches up their interpretations to coincide with the Bible. Either God's Word is correct, or man's wisdom is correct.

The pieces of the jigsaw puzzle are slowly falling into place. What caused the global flood—the canopy of salt water—is the very same source that kept the planet in ambient temperatures for maximum vegetation growth and prevented polar ice caps from forming. The extra abundant vegetation produced a greater percentage of oxygen, which allowed life to thrive. Hence, human beings and other creatures lived 900+ years before the Flood. This canopy also increased atmospheric pressure on the inhabitants of the earth, which created a large buoyant force that reduced the net effect of gravity, and it allowed dinosaurs to grow extremely large with porous bones and to live extremely long lives because of increased oxygen. But the canopy was lost to the global flood. Notice in Gen. 7:12 that it says, "The rain fell upon the earth for 40 days and 40 nights." Since water was stored up above the atmosphere beyond the earth, and the writer (Moses) was describing water coming down upon the earth, this makes sense that he wrote that "rain fell upon the earth." He doesn't say that rain fell upon a region, but upon the earth. This global rain and subsequent flood plunged the earth into a cold spell called the glacial age some 4,400 years ago.

Chapter Summary: The polar ice caps and the glacial age formed as a result of 40 days and nights of rain and no sunlight (Gen. 7), with the waters being absorbed by the polar regions over the following 12 months. Before the global flood, the canopy of salt water hovered around the

atmosphere and created ambient global temperatures, preventing polar ice caps and deserts from forming. The glacial age began around 2,400 BC.

Group Discussion:

1. What convinced you that the polar ice came suddenly and quickly? Was it the tropical forest buried beneath the ice? Was it the woolly mammoths frozen with undigested food in them? Was it the WW2 airplane buried deep in the ice? Or was it the Bible because it first mentions cold and winter after the Flood?

2. The old ice age hypothesis proposed by evolutionary glaciologists is faith based and is a theory with problems in the dating method. Its speculative dates are merely established to coincide with an old earth hypothesis. Do you have the same level of faith in the Bible as evolutionists have in their beliefs?

If you have comments or questions to the author, email them to: Lawrence@creationministry.org.

END.

www.ingramcontent.com/pod-product-compliance
Lightning Source LLC
Chambersburg PA
CBHW081017040426
42444CB00014B/3248